The V

EXTRA
DIMEN
-SIONS

Volume 1

Visualizing the Fourth Dimension,
Higher-Dimensional Polytopes,
and Curved Hypersurfaces

Chris McMullen, Ph.D.

The Visual Guide to Extra Dimensions

Volume 1: Visualizing the Fourth Dimension, Higher-Dimensional Polytopes, and Curved Hypersurfaces

www.faculty.lsmsa.edu/CMcMullen

Custom Books

Nonfiction / science / physics

Nonfiction / science / mathematics / geometry

ISBN: 1438298927

EAN-13: 9781438298924

Volume 1
Visualizing the Fourth Dimension, Higher-
Dimensional Polytopes, and Curved Hypersurfaces

Volume 2
The Physics of the Fourth and Higher Dimensions,
Compactification, and Current and Upcoming
Experiments to Detect Extra Dimensions

The Visual Guide to Extra Dimensions

Volume 1

Visualizing the Fourth Dimension, Higher-Dimensional Polytopes, and Curved Hypersurfaces

Chris McMullen, Ph.D.

Contents

Introduction

My fascination with extra dimensions began when I was in high school and encountered one of Rudy Rucker's books on the fourth dimension during one of my trips to the math and science shelves of a bookstore. I remember staying up into the wee hours contemplating the ana and kata directions, drawing hypercubes and hyperspheres, and solving the puzzles from Rudy Rucker's book.

I was fortunate to combine my interest in extra dimensions with my expertise in physics during my doctoral studies at Oklahoma State University: Some papers in the 1990's had stimulated interest in large string-inspired extra dimensions in the particle physics community. I earned my Ph.D. in physics when an explosion of professional research was published in this field, and have continued to publish papers on the collider phenomenology of large extra dimensions ever since.

Recalling my appreciation for Rudy Rucker's works, I wanted to make my own humble contribution to readers with an interest or background in math or physics who are curious about extra dimensions. I hope that my work will engage the interest of some readers who share my curiosity for the fourth and higher dimensions.

This first volume is dedicated toward a geometric extra dimension very much like the three known dimensions. It begins by surveying interesting features of 1D and 2D worlds, since we can understand many higher-dimensional concepts through analogies in the lower dimensions. Some common objections to studies of a fourth dimension are addressed, especially to convey the message that times have changed: There is good motivation for the possibility of superstring-inspired extra dimensions that are much larger than originally thought – large enough that current and upcoming experimental searches are underway with prospects for detecting them.

One of the main goals of the first volume is to develop techniques for visualizing extra dimensions; numerous illustrations, many of which are novel, aim to aid in this process. Another goal is to thoroughly analyze some fundamental 4D objects, especially the tesseract and the glome. The final chapter of this volume looks at a hypothetical hyperuniverse, in which these visualization techniques and basic geometric objects are applied. This chapter serves as a useful bridge between the higher-dimensional geometry considered in the first volume and the higher-dimensional physics developed in the second volume.

This book is primarily conceptual, for the benefit of readers who do not have a strong background in the mathematics of particle physics or superstring theory, yet there is a mathematical component to this book, since it is anticipated that most readers will have some interest or background in mathematics. It will be desirable to have a good handle on the fundamentals of algebra, geometry, and graphing. Where it is deemed useful to discuss more advanced mathematics, higher-level concepts are developed from these starting levels.

Nonetheless, more emphasis is placed on the concepts than on the math, since this is intended to be an enjoyable book on an interesting topic, which should not read like a textbook. This book is very detailed and technical conceptually in an effort to help stimulate and engage the interest of such mathematically-minded readers.

Several illustrations are intended to challenge readers visually. For example, several illustrations combine together to show a hypercube unfold, the various 3D cross sections of a hypercube are drawn, the 3D projection is depicted for a rotating hypercube, higher-dimensional cylinders and tori are graphed, and a higher-dimensional staircase is drawn. Puzzles scattered throughout the book offer additional challenges.

The second volume looks at higher-dimensional mathematics, higher-dimensional force laws related to Gauss's law, the important issue of compactification, current and upcoming experimental searches for extra dimensions, and a little background in spacetime, quantum mechanics, and string theory.

May you enjoy the book as much as I enjoyed reading my first book on extra dimensions.

Dedications

I would like to thank the many teachers who inspired my interest in math and science, my passion for teaching, and many of the good qualities that I strive for in my teaching, research, mentorship, and life. I particularly appreciate those teachers who were excellent motivators and mentors and those whose courses were very challenging, rigorous, and thought-provoking. One of my favorites is Dr. Robert Chianese, who taught a fascinating course on Cold War Literature – so much so that I have since re-read many of the assigned readings, putting forth much more effort after the credit had already been earned. I had the good fortune of taking a few courses with Dr. Duane Doty, who is a superb mentor, motivator, and teacher. My Master's thesis, under the guidance of Dr. Miroslav Peric, immensely helped to develop my confidence in challenging myself with research. Teachers such as Dr. A.C. Cadavid, Dr. Peter Collas, and Dr. Robert Park helped to instill my interest in rigorous mathematics, and inspired the high level of mathematics that I strive to incorporate into the courses that I teach. Dr. K.S. Babu is the ideal model for a command of knowledge in one's area of expertise. I am very grateful for the opportunities that I have had to collaborate with my mentor, Dr. S. Nandi, on multiple research projects, in addition to being a student in the excellent courses that he taught. I would also like to mention my high school geometry teacher, Mr. Ratkovic, who is not only an exceptional teacher, but who identified gifted students in his classes and found ways to engage their interest and challenge their minds while still teaching to the rest of the class. I appreciate that so much that it is one of my primary reasons that I accepted my current position at the Louisiana School for Math, Science, and the Arts, and it serves as a strong source of my motivation to challenge young minds in math and physics.

I must also thank my family for their encouragement and always believing in my ability, including my mom, dad, grandma, and aunt. I am also very grateful for the invaluable support of my wife, whose Master's thesis was also on the subject of collider phenomenology of large extra dimensions.

I dedicate this book to all those who have had a positive impact on my life.

0 The Known Dimensions

Ironically, you pick up this book on extra dimensions, and it begins by discussing the lower dimensions. Of course, there are reasons for this. For one, it is necessary to develop a good idea of what a dimension is and how to determine the dimensionality of a space in order to understand what is meant by an *extra* dimension. For another, there are many concepts in higher dimensions that are analogous to simpler concepts in the lower dimensions, and these analogies help to provide insight into the higher dimensions. The tesseract, for example, which is a four-dimensional generalization of the cube, can be better visualized and understood by examining relationships and patterns between lines, squares, and cubes. But even without extra dimensions in mind, you may find that some aspects of the lower dimensions are actually quite fascinating. For example, it turns out that buildings and beings in a two-dimensional world would inherently be much different from what we are accustomed to drawing on a sheet of paper.

0.0 What Is a Dimension?

In order to convey precisely what is meant by lower or extra dimensions, let us begin by examining just what a dimension is.

The English word *dimension* derives from the Latin word *dimensio*, meaning extent. The Latin word *dimensio* is itself a combination of the prefix *dis-* and the Latin infinitive *metiri*, meaning to measure. Curiously, in this case the prefix *dis-* does not serve as a negation – as in *dis*prove (otherwise, *dis-* combined with *metiri* would mean to not measure). Instead, the prefix *dis-* acts as an intensive – i.e. it adds emphasis to *metiri*. The prefix *dis-* has a similar effect in *dis*turb, where *turba* means confusion, and in *dis*annul, which means to nullify.

While the etymology shows how the term evolved, it does not serve as precise definition. Here is what the term *dimension* means:

A dimension is a measure of extent.

The most common connotation is spatial extent, as in the dimensions of a rectangular block – i.e. its length, width, and depth. A dimension can also refer to temporal extent, as time is a measure of the extent of an event or the duration between two events. Physicists, in

11

fact, adopt seven fundamental dimensions – length, time, mass, temperature, electric current, luminous intensity, and mole number (or amount of substance) – from which the dimensions of any other physical quantity can be derived – e.g. velocity has dimensions of length divided by time. Occasionally, the definition may even apply to a non-physical attribute – e.g. a one-dimensional character. However, we have a much more limited usage in mind for our purposes:

In this book, the word *dimension* will be used to imply spatial extent, unless noted otherwise.

We will also want to distinguish between the similar terms, *dimension* and *dimensionality*. The dimensionality of a space or an object refers to the number of independent directions into which it extends. Here are a few examples of objects with different dimensionality:

- A single point, infinitesimal in size, has no spatial extent. Points are zero-dimensional because they extend in zero directions.
- A line segment, extending along a single direction, is one-dimensional (1D, for short).
- A rectangle is 2D, as it extends along two independent directions – length and width.
- A 3D rectangular block has a third independent direction – depth.

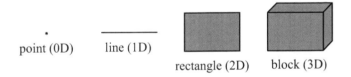

point (0D)   line (1D)

rectangle (2D)   block (3D)

It is easiest to count the number of dimensions of a rectangular object or space. In order to determine the dimensionality of curved spaces, it will be important to understand what is meant by an independent direction. Following are some examples to help illustrate this concept:

- North and east are independent. When sailing due north, a boat does not move west or east at all. Conversely, sailing due east does not involve any north or south motion.
- Longitude and latitude are similarly independent. Longitudes are circles intersecting the poles, while latitudes are circles parallel to the equatorial plane. The latitude of an airplane does not change if it flies along a latitude, nor is its longitude altered when

traversing a longitude. These two sets of circles correspond to two independent directions of motion.

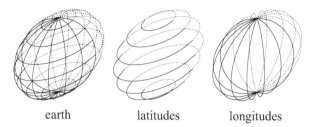

earth   latitudes   longitudes

- North and south are not independent. Walking south can be thought of as walking in the negative north direction. Put another way, when you walk south, you are getting less north. If you travel north and then south, you are experiencing 1D motion. In fact, the only way to change direction in 1D is to reverse direction.
- Northeast is not independent of north and east: When driving northeast, a car is simultaneously moving partially north and partially east.
- However, northeast and southeast are independent. These can be seen to be equivalent to north and east by rotating a map 45°.
- The length, width, and depth of a rectangular block are independent.

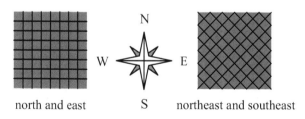

north and east   S   northeast and southeast

    The zeroth dimension is an infinitesimal mathematical point. Ordinary human activities that are largely 0D include sitting, sleeping, and occupying a prison cell. This latter activity is often referred to as "doing time," since space is severely limited. Insofar as time is a temporal dimension, and an integral part of spacetime, these activities are 0D in space, yet 1D in time. In order to eliminate the temporal dimension, one must freeze time – as in a snapshot.

    The zeroth dimension is trivial – it would be pointless (literally!) to give too much thought to it. Actually, "thought" is pretty much the only activity that one could do in 0D, which is probably why Edwin Abbott called it Thoughtland [A1]. A 0D universe would not be physical, but philosophical. So, a 1D space is the lowest dimensionality that actually has physical extent (unless you really want to complicate matters and contemplate fractional dimensions). Let us open up the lower dimensions one at a time so that we may progressively see the extra freedom that is earned in doing so and make comparisons that will be useful later.

0.1 The First Dimension

The first dimension features extent along just one independent direction. A 1D object could be linear, but we shall learn that it could also be curved.

An object residing in the first dimension has just one degree of freedom: It can only move forward or backward. Many ordinary human activities in our universe are largely 1D: An elevator moves only up or down, a roller coaster winds its way along a track, a raft follows the course of a river, and a country highway provides a transportation route.

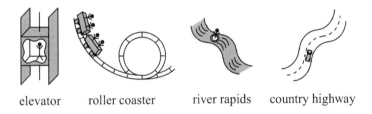

elevator   roller coaster   river rapids   country highway

Notice that winding along a curved path corresponds to largely 1D motion. It might seem that a roller coaster ride is a 3D experience, but the roller coaster is actually confined to a preordained 1D path: The cars may only move forward or backward (or not move at all) as they are constrained to follow the course of the track. The cars have a single degree of freedom. Thus, a roller coaster ride is effectively 1D, even though the track may wind its way through a 2D or 3D Euclidean space.

A Euclidean space is "flat" – like a line or a plane; a curved space is said to be non-Euclidean. A Euclidean space is naturally described by rectangular (or Cartesian) coordinates – x and y in 2D, and x, y, and z in 3D, for example. If you are stuck in a Euclidean mindset, you might have a bias here – i.e. you might want to think of parabolas, circles, and other curves as 2D (or higher, as in the case of a helix). In math, you learn that you need two Cartesian coordinates to describe the path of a parabola (e.g. $y = x^2$) or circle (e.g. $x^2 + y^2 = 1$). However, working with a set of appropriate curvilinear coordinates, it is possible to describe such paths with a single coordinate that indicates, conceptually, how far forward or backward a point is along the path. For example, you only need to specify an angle (relative to some reference point) to know the exact location of any point on a circle; the two Cartesian coordinates x and y are not actually independent – knowing one, you can calculate the other (up to a sign, since for any value of x there are generally two values of y, one positive and one negative).

If you do have a Euclidean mindset, what you are actually thinking is that a 1D curve winds its way through a 2D (or higher) Euclidean space. As a result, you may be wondering why it makes sense to say that a curve is 1D, when you must have a 2D (or higher) Euclidean space to "house" the curve. A practical reason is that it is simplest to work with the number of independent coordinates, which may be less than the number of Cartesian coordinates needed to describe the object or space; a practical reason has to do with compactification. We will consider compactification in more detail in Chapter 8. For now, it will suffice to have a basic sense of what it means for a space to be *compact*.

As a simple illustration of compactification, consider a long, thin, straight wire. If it has negligible thickness, then the wire is essentially 1D. Slip a bead over the wire, and the bead can only change location by moving forward or backward along the wire. Now imagine bending the wire into the shape of a circle. The bead can still only move forward or backward along the wire, so we also classify this circular wire as 1D. The straight wire represents a 1D Euclidean space, whereas the circular wire represents a compact dimension. The wire is said to have been *compactified*. Any extra dimensions in our universe may be compactified in a similar manner.

In contrast, if you cut out a circle in a sheet of paper, this circle is 2D because it includes the region inside. An ant crawling on the circular cutout experiences two degrees of freedom. The dimensionality depends upon whether or not the interior region is a physically accessible part of the space. When we refer to a circle, we will mean just its circumference unless stated otherwise. Similarly, when we refer to a sphere, we will mean just its surface, and we will instead use the term *ball* to indicate a sphere that includes the interior region.

15

A curved 1D universe could be closed, like an ellipse or circle, or it could be open, like a parabola or hyperbola. In a 1D universe that is open, after moving forward, it is necessary to move backward to return to a starting point. However, in a 1D universe that is closed, it is possible to travel all the way around the world, moving only forward, to return to a starting position. In this case, the motion is periodic.

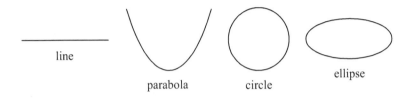

line

parabola   circle   ellipse

A 1D universe need not be continuous. A black hole, for example, would be a discontinuity in a 1D universe. In a noncompact 1D universe, it is not possible to travel around a black hole to reach a destination on the other side; the black hole divides such a 1D universe into two separate regions.

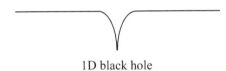

1D black hole

It is also possible for multiple 1D worlds to connect in some way. For example, two 1D worlds can intersect at a point, or two parallel 1D worlds can be joined with a bridge (aka wormhole). For a 1D universe that curves through 2D (or higher) Euclidean space, such a bridge could provide a shortcut.

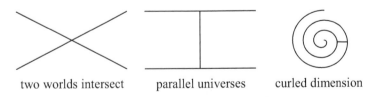

two worlds intersect   parallel universes   curled dimension

So far we have considered possible structures of the universe itself. Let us now look inside the 1D universe, and contemplate what types of objects may exist there and how they may interact. In doing so, here is a fundamental question to consider: Are the objects in the 1D universe necessarily 1D themselves? Intuitively, it might seem that the objects in a 1D universe would inherently be 1D, objects in a 2D universe would be 2D, etc. However, it is also possible for 0D objects to reside in a 1D world, 0D and 1D objects to reside in a 2D world, etc. In fact, it is more than possible – close examination of our universe reveals that all objects in our universe are composed of elementary particles (like the quarks that make up protons and neutrons in the nucleus of atoms and the electrons that reside in electron clouds outside the nuclei). If they are truly elementary (i.e. they can not be subdivided), then they may be 0D points. If they have structure, then it seems reasonable that they could be split up into smaller components, and hence would not be elementary. A pure mathematical point has no structure, so elementary particles are thought to be pointlike. Thus, although our universe is at least 3D, all objects in our universe are apparently composed of 0D particles. By analogy then, in a 1D world, there could be elementary 0D particles that bind together microscopically to form macroscopic 1D objects.

0D objects   1D objects   0D and 1D objects

Perhaps you are wondering whether elementary particles are actually strings, rather than points. Or if you know some quantum mechanics, you may know that elementary particles actually exhibit wavelike characteristics, in addition to particle-like properties, and so are wondering how precise it is to state that an elementary particle is pointlike. These are good points to bear in mind. Some principles of quantum mechanics will be examined in Chapter 10, especially as they relate to superstring theory.

It seems that objects would have very little freedom in a 1D universe. Any object residing in a 1D world would be confined to the region between its nearest neighbors. Otherwise, a single relatively immovable point would act as a wall, and two such points would create a prison. However, there is a way around this. The ability of particles to pass through other particles would immensely increase freedom of motion in 1D. If you think of this as ghosts passing through walls, you might laugh, but it could be more like quantum tunneling – a phenomenon actually observed in our universe, although in 1D it would not just be a matter of an electron passing through a forbidden domain, but a particle passing directly through another particle. On the other hand, a 1D universe may not resemble our universe. In a 1D world, the ability of particles to pass through other particles – perhaps with some probability of passing through, at least – would serve a very useful purpose.

17

imprisoned

Another feature to consider is communication – i.e. how objects interact with one another. In our universe, there are four fundamental interactions: Particles can interact through a gravitational force, an electromagnetic force, a strong nuclear force, or a weak nuclear force. It is conceivable that a 1D universe may be governed by a much different set of fundamental interactions, so how one object would "see" another object in 1D may involve something much different from what we perceive as light (and thus may not behave like an oscillating electromagnetic field). Nonetheless, one method by which particles may communicate is to send out a mediator – like a photon, a particle of light. By exchanging mediators, particles exert a force on one another, which affects their motion.

Imagine something like a 1D lightbulb, and let us call the mediator "light" while allowing for the possibility that it may be fundamentally different from what we call light. If the nearest object on either side of the lightbulb absorbs the light, then only the nearest neighbors can "see" the light. In order to illuminate a greater region, the surrounding objects must be transparent. The surrounding objects might temporarily absorb the message, and then re-radiate it in the same direction – passing it along as schoolmates might distribute a note during class. A problem that results is distinguishing between numerous light sources on either side and determining distances. An interesting feature possible in a curved 1D universe is that distant objects can be illuminated if light propagates into two (or more) dimensions, even if the inhabitants of the 1D world are themselves are confined to 1D motion.

Objects residing in a 1D world can interact by exchanging mediators (like sending and receiving particles of light), but there are additional modes of communication available. For example, there is possibility of action at a distance, in which case in-between objects are "skipped" in more distant interactions. Alternatively, distant objects can affect one another indirectly by colliding with their nearest neighbors, which would cause subsequent collisions in a domino effect.

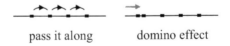

pass it along   domino effect

There is one important feature of our universe which may intuitively seem impossible in 1D – orbits. At least, it would not be possible for one object to travel in a closed orbit, like a circle or ellipse, around another object in the same 1D space. In our universe, orbits are very fundamental to solar systems at the macroscopic level and to atoms at the microscopic level. Let us consider what atoms or solar systems may be like in a 1D universe.

One simple idea is that the nucleus of an atom would interact with its surrounding particles as if the nucleus and surrounding particles were all connected by springs. In this case, the "orbits" are simply back-and-forth motions, or oscillations. Atoms could bind together with other atoms to form 1D chains, and a large number of such chains could form macroscopic objects. Planets could similarly oscillate back-and-forth on either side of a star. Notice that in 1D, planets feature a very large interior, but just two pointlike surfaces – one on each side. Similarly, stars emit light that can be seen from just two directions.

Another possibility is that the surrounding particles pass through the nucleus, traveling back-and-forth from one side to the other. In this case, the particles travel back-and-forth through the nucleus. Allowing for discontinuous jumps – like hyperspace for a spaceship in a science fiction book, or quantum tunneling at the microscopic level – is another means of creating some type of orbit in 1D.

hidden springs

hyperspace orbit

Here is one more 1D analogy with our universe, and then the rest will be left to your imagination. Just like an open cosmological question in our universe, a 1D universe could expand or contract. The "stars" of a 1D universe would become farther apart from one another if the universe were expanding. Even in a curved 1D universe, the effects of expansion can be achieved: For a circular universe, increasing the radius causes the "stars" to become farther apart.

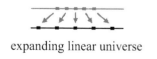

expanding linear universe

expanding circular universe

Even though it seems plausible to make a 1D universe that has some analogies with our universe – especially, by allowing particles to pass through one another to some degree – living in such a world would be very confining. Next, we will open up the second dimension and consider the impact of this extra degree of freedom.

0.2 The Second Dimension

The second dimension might seem familiar, since we have experience drawing objects on a sheet of paper. However, these drawings are 2D representations of 3D objects, and we will see that they may differ greatly from objects in a 2D world. It turns out that many features of a 2D universe would be much different from life in our universe.

Barring any constraints, objects in a 2D universe have two degrees of freedom. In addition to forward/backward, there is left/right. Alternatively, the two independent directions could be characterized as north/south and east/west. Ordinary human activities in our universe that are largely 2D include: walking in an open space (as opposed to, say, hiking along a trail or walking along a sidewalk), sailing, and mowing the lawn.

Imposing a constraint may reduce the degrees of freedom. For example, suppose an object is allowed to move in combinations of north/south and east/west, but must always be 10 m from a particular tree. The result is that the object can only move along the circumference of a circle, with a single degree of freedom – namely, clockwise/counterclockwise. The motion of this object is effectively 1D. However, if the constraint were modified to state that the object cannot go further than 10 m from a particular tree, this corresponds to a 2D space. In this case, the object can move in the interior of the circle, and so has two degrees of freedom anywhere within the circle.

motion along a circle (1D)   motion within a circle (2D)

Imagine what a 2D universe might be like. A 2D universe could be flat, like a plane, or curved, as in the surface of a sphere or a cylinder. Again, the terms *sphere* and *cylinder* refer just to the surfaces of these objects. Just as an ant crawling on a tennis ball or soda can has two degrees of freedom, these curved surfaces are considered to be 2D.

plane   sphere   cylinder

Compared to 3D, it is easier to visualize features of 2D universes such as black holes, where the surface is pinched down to a single point. Most representations of black holes in our universe are really what 2D black holes would look like because it is much more difficult to draw a 3D black hole properly. Similarly, it is easier to imagine two parallel 2D universes than parallel 3D universes – since in the former case a universe may look like a plane rather than a 3D space. Two parallel 2D worlds could connect in some way, such as a simple flat bridge or a curved wormhole. Two nonparallel 2D worlds could connect by intersecting. One of the dimensions could be curled up – in this case the 2D universe would look like a rolled-up carpet.

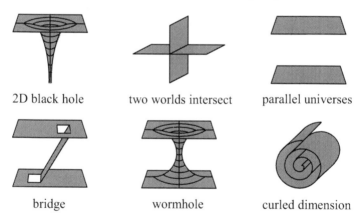

2D black hole   two worlds intersect   parallel universes

bridge   wormhole   curled dimension

In 2D, light can radiate outwards from a source, such as a star or a lightbulb; it cannot be blocked by a single pointlike object as it can in 1D. However, unless the photons are not confined to the same 2D surface as the particles, the physics of electromagnetism (light) would be somewhat different. If the physics is analogous to that of our universe, then Maxwell's equations, which govern the behavior of light, would require that Coulomb's law for the attraction or repulsion between electric charges be modified from an inverse-square law to just an inverse (no square); there would be a similar modification of the inverse-square law of gravity, too. Since these laws are fundamental to atomic and solar system structures, including allowed orbits, this change in the power laws would cause the nature of the 2D universe to be significantly different from our universe.

On the other hand, it would be possible to make a model of a 2D universe where these two power laws remain inverse-square laws, but as a consequence there would be a significant change in the underlying physics: Demanding that Coulomb's law be an inverse-square law in 2D would alter the nature of light, as it requires an alternative to the 2D analog of Maxwell's equations. We will return to this issue Chapter 7, when we introduce Gauss's law.

In contrast to 1D, it is possible to have closed orbits in 2D – e.g. electrons orbiting a positively charged nucleus or planets orbiting a star. However, atomic structure and bonding in 2D would be fundamentally different from that of our universe.

Now think what it would be like to live in a 2D universe. Simple machines, from which all machinery is built upon, would be much different in 2D from our universe. Let us consider each of these in turn.

A circle serves as a wheel, useful for transportation or generating electricity in our universe, but in 2D it would not be a trivial matter to connect the wheel to something, such as a bicycle, without drilling a hole straight through the radius of the wheel – in which case, the axle would prevent the wheel from rotating. In 3D, the axle is perpendicular to the wheel, passing through its center without destroying the use of the wheel, but the axle simply cannot be perpendicular to the wheel in 2D without passing through its circumference of the wheel; it cannot simply be beside the wheel.

wheel　　　　axle prevents rotation

A nail is a common example of a wedge in our universe, fundamental to construction. However, in 2D, upon attempting to hammer a nail through a board to fasten it to another object, the board splits in half! With nails utterly useless in joining two objects together, glue would be a precious commodity. The nail instead serves the purpose of a saw in 2D. Also, note that the head of the nail would not be a circle.

board and nail   board split into two pieces

A line in 2D can serve the same purpose as an inclined plane in 3D. This would be useful for making ramps. In 3D, the screw is a very useful combination of the wedge and the inclined plane: A screw is formed by winding an incline around a nail. It is not only useful for construction, but the Archimedean screw is very useful in raising large amounts of water to greater heights. However, the screw is a 3D object, which is not possible in 2D.

Pulleys and gears suffer from the fundamental problem of the wheel in 2D – how to connect the axle to the wheel without preventing the wheel from rotating. A pulley, at least, could be supported from the bottom, with gravity pulling down, provided that there is not enough friction between the wheel and the support, or between the wheel and cord, to pull the wheel off the support; some grease can help.

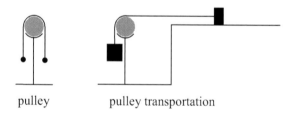

pulley   pulley transportation

Nails and screws are not the only fasteners that are problems in 2D: It is not possible to tie a knot with a rope; the two ends can not pass around one another without splitting the rope. Yet, there are alternatives to glue and tape. For example, objects can be formed with male and female fittings that snap into place. Nails, actually, can serve a purpose other than splitting objects in half: If the nail is only hammered partially into a wall, it can still serve to hang something.

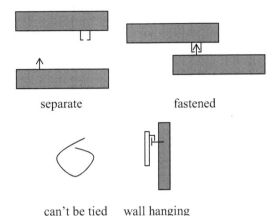

separate   fastened

can't be tied   wall hanging

A 2D being could use a wheel for transportation by running on top of the wheel the same way that a logger spins a log on a river. An improvement on this is to place multiple wheels inside a belt – like a tank. A wheel, or set of wheels, positioned behind an object could push the object forward provided that there is not too much friction between the object and the ground. Magnets could be used to attract wheels to an object to form a 2D vehicle. In this case, lubricant could allow a separate rim of the wheel to rotate while the central circular magnet remains in a fixed orientation. Pulleys and gears can also work in a similar fashion. Notice that in 2D a single wheel is perfectly balanced – it can't fall over.

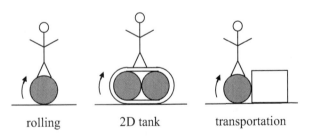

rolling   2D tank   transportation

Now we will contemplate what it may be like for a 2D being to live in a 2D world. Assuming that there is gravity, a tree, or even a vertical line, serves as a wall in 2D: It is not possible to walk around a tree; in order to pass it, it is necessary to climb over the tree. Even if the tree is chopped down, it is still necessary to walk over the pieces or pass them overhead. Trying to dig a tunnel under the tree, the tunnel would collapse due to lack of support. A square, or any polygon or closed curve, serves as a prison in 2D.

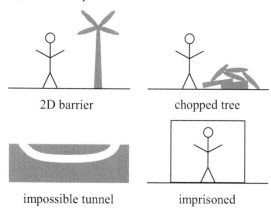

2D barrier   chopped tree

impossible tunnel   imprisoned

Designing a 2D house, there is the important issue of how to get in and out. Doors cannot function in the usual way, being hinged along one edge; instead, they must slide or be hinged at a point. For a simple house with one door in the front or back, the house would tip over if the door were opened! A small entry way with doors on each side could prevent this, provided that the first door is shut before the second is opened. Alternatively, there could be a staircase or ladder on one side of the house and a hatch on the roof.

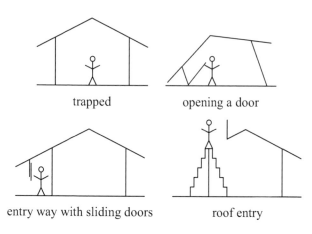

trapped   opening a door

entry way with sliding doors   roof entry

In 2D, a pair of parallel lines can serve as a pipe. The problem with using a pipe is to prevent it from collapsing since the two sides of the pipe cannot be connected without blocking the pipe. An

25

underground pipe would tend to collapse as easily as a tunnel. Telephone cables pose a different problem. The cables can be thrown over telephone poles – no need to fasten them, as they will not fall off – but the poles create barriers; they serve as walls. Trying to climb over the telephone pole, the cable serves as a ceiling. By making telephone poles from two pieces, where the top piece rests inside the bottom piece, it would be possible to lift the top piece to walk through the telephone pole. This could become tedious during a long walk.

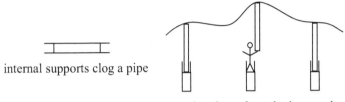

internal supports clog a pipe

passing through a telephone pole

There could be ponds, lakes, and oceans of liquid – not quite the same as water, since the physical and chemical properties would be different in 2D. Assuming that there is gravity, it would not be possible to walk around a pond – to reach the other side requires crossing it. Rain falling down a mountain in the form of a river produces a flood – there is no escaping a river of water. Trees, houses, and other barriers act as dams, until they give way or the water level rises above their heights.

liquid barrier

An arc would serve as a boat in 2D. Rowing requires using an oar at either the bow or the stern of the boat; the usual meanings of port and starboard are lost. A sail would simply be a long, thin strip of cloth. The mast would block wind from reaching the sail from the fore or aft direction – whichever is opposite the sail – unless the mast supports two sails. An airplane cannot have wings in the usual way, neither could it have a propeller; nor would the usual horizontal rotor be possible for a helicopter. A 1D curve could contain the gas of a hot air balloon. The ropes would need to be fastened by some means other than tying, and the ropes would need to be undone and refastened in order to allow passengers to enter or exit.

rowboat   sailboat   hot air balloon

Now let us turn our attention to the 2D being itself. An interesting feature is that a digestive track would split the body in half. A single orifice for digestion – like an earthworm – would solve this problem. An alternative is a passageway that unseals and reseals as necessary. There is a similar problem with a respiratory system in which air passes through both the nose and the mouth. Birds could not have wings at the sides, as in 3D; the wings must be above, below, ahead, or behind.

Imagine a 2D being similar to a human, in a 2D world with gravity. It would seem strange for the eyes, nose, and mouth to lie on one side of the face. It would be very useful to have an eye on the back of the head, especially since it is not possible to turn around without turning up-side-down. Walking backward would be awkward if both feet faced forward. Even walking forward is difficult, since it is not possible to put one foot in front of the other. Knees that bend both ways would make it possible to sit on a chair in the front or back. During a night's sleep, to switch from lying face down to face up would be cumbersome: It requires sitting up, falling over, and stretching the legs, in addition to adjusting the sheet and transferring the pillow to where the feet had been.

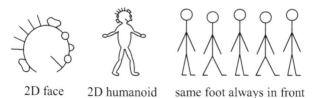

2D face   2D humanoid   same foot always in front

tossing and turning at night

In a 2D world with gravity, it is not possible to pass "around" another person. Passing requires one person to pass over another – reminiscent of children playing leap frog. The rear hand is not in a position to help the front hand with lifting. If all people face the same direction – say, to the right – then kissing would not be as simple as it is in our universe. It may be practical for males to face one direction and females to face the opposite direction. It is not possible to wrap arms around someone in a hug. Hands could not clasp the same way in a handshake as they can in 3D; the next best thing would be to clasp fingers.

passing through

Another interesting notion is 2D clothing. If there is gravity, 2D "shorts" would fall apart, as they would consist of separate pieces of cloth. Even socks would fall right off. Pants could consist of a single piece of cloth provided that they included feet like a toddler's pajamas, but it would not be possible to wear a belt or suspenders to hold them up. The next step is overalls with built in shoes and gloves, but these would easily slip off, too. A shirt poses another problem: It must completely cover the head. Simply poking an eyehole into a shirt splits it into two pieces. Transparent material could be used at eye level, but poking breathing holes in the material poses yet another problem. Ties would need to clip on.

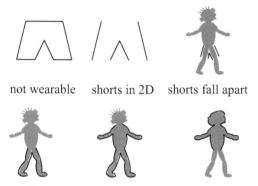

not wearable   shorts in 2D   shorts fall apart

pants fall off   overalls slip off   shirt covers face

Notice that we have been visualizing the second dimension from the perspective of a third dimension – i.e. we are viewing possible 2D worlds from a point of view that is not possible for a 2D being living in such a world. Look at the eye of a 2D humanoid and imagine what he or she sees. In 3D, light reaching the eyes results in a 2D image in the mind. By analogy, in 2D, light reaching the eyes would result in a 1D image. Drawings, photographs, and television screens would also be 1D. A tree, for example, would just look like a line; however, some points may appear brighter than others, and different objects may be distinguished by their color. Two eyes on the same side of the head would aid in depth perception.

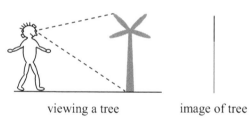

viewing a tree   image of tree

There would be much more freedom in 2D if there were no gravity. It would be possible to move around people, trees, and other objects. However, without gravity pulling down, objects would float through the 2D space, provided that Newton's concept of inertia holds in the 2D world. In this case, 2D beings would float, rather than walk. This is a fundamental difference between an ant in our universe crawling on a sheet of paper and a truly 2D being wandering through a 2D world.

29

Adding one more degree of freedom brings us to the third dimension, an experience with which we are most familiar.

0.3 The Third Dimension

Our universe *appears* to be 3D – with emphasis on *appears* since there are reasons from theoretical physics to suspect that there may be extra dimensions, and there are active experimental searches underway to detect them. We will later return to such questions as how to reconcile the possible existence of extra dimensions with the observation that structures and motion in our universe appear to be 3D, and how and why physicists are searching for extra dimensions.

Aside from any constraints, objects in a 3D world have three degrees of freedom: forward/backward, left/right, and up/down. Alternatively, the three degrees of freedom may be characterized as longitude, latitude, and altitude (spherical coordinates). Ordinary human activities in our universe where the full 3D freedom is experienced include scuba diving, hang gliding, and playing on a jungle gym. Most day-to-day activities are largely 0D-2D.

Since we are all familiar with 3D motion, here we will focus on how we perceive the three evident dimensions of our universe. This will provide some insight into how to visualize the higher dimensions.

3D eyes have a 2D surface; light entering the eye leaves a 2D image in the mind. However, macroscopic beings gain a perception of depth through experience, which is aided by having two eyes. The two eyes actually see slightly different images; the mind merges these together. This can be tested by exploring parallax.

The phenomenon of parallax can be experienced by placing one finger about one foot before the eyes. First, notice that the finger and distant objects cannot both be in focus simultaneously: When the finger is in focus, the background is blurry, and vice-versa. With the background in focus, close one eye, leaving the other open. Then switch eyes. The finger appears to shift position. Now move the finger further from the eyes, and try it again. Notice that the amount of shift, or parallax, depends on the distance between the finger and the eyes.

3D objects can be represented with 2D diagrams. However, using the two independent directions available on a sheet of paper to illustrate 3D objects comes at a cost – some ambiguity in gauging depth. Horizontal and vertical are two independent directions on a sheet of paper. A third direction independent of these is perpendicular to the sheet of paper – i.e. in front of or behind the sheet of paper. It is not possible, of course, to make pencil marks, say, 2 cm in front of or behind the sheet of paper. Instead, the diagonal direction is utilized to illustrate the third dimension. The ambiguity in interpreting 2D drawings of 3D objects arises due to the fact that diagonal is not independent of horizontal and vertical.

Do line segments $\overline{AE}$, $\overline{CF}$, and $\overline{DG}$ go into or out of the page?

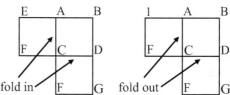

two ways to fold this pattern into the shape above

One way to draw a 3D object is to map out its coordinates on a 3D graph. The rectangular, or Cartesian, coordinates x, y, and z can be used to make 3D graphs. If x is to the right and y is upward, then z comes out of the page. The coordinates $(3,2,1)$, for example, mean 3 units along x, 2 units along y, and 1 unit along z. The point $(3,2,1)$ really lies 1 unit in front of the page above the point $(3,2)$ in the xy plane; however, since it is not possible to draw the true point, the z-coordinate is drawn one unit diagonally down to the left.

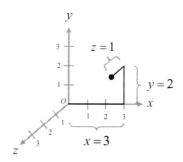

locating the point $(3,2,1)$

31

Given the Cartesian coordinates of a point, there is a definite way to locate the point on a 3D graph. However, the converse is not true: Looking at a point on a 3D graph, there is some ambiguity in interpreting its Cartesian coordinates. This ambiguous interpretation represents some information that is lost in the process of illustrating a 3D object with a 2D sheet of paper. It will be important to keep in mind that there is even more ambiguity in representing the fourth dimension. Nonetheless, 4D objects can be still be visualized and understood, although more time and effort is required to understand them well.

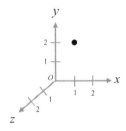

What are the coordinates of this point?

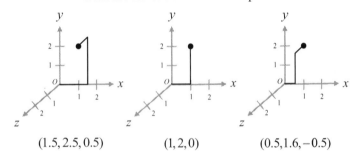

$$(1.5, 2.5, 0.5) \qquad (1, 2, 0) \qquad (0.5, 1.6, -0.5)$$

Our universe appears to be 3D to macroscopic beings. Rather, it appears to be $(3+1)$-D: three spatial dimensions plus one temporal dimension. Macroscopic beings experience three independent directions of motion: They can move north/south, east/west, and up/down. It appears to be impossible to move in a fourth direction that is independent from these. Thus, our universe is thought to consist of three spatial dimensions. Nonetheless, it is possible that there are extra spatial dimensions that macroscopic beings cannot detect directly. The universe may turn out to be $(N+1)$-D, where $N > 3$. Henceforth, we shall consider such possible extra dimensions in detail.

Further Reading

Edwin Abbott's *Flatland* [A1] is a classic 4D novel about a 2D being's journey through the third and fourth dimensions. The modern follow-up to this work is Ian Stewart's *Flatterland* [A2]. Rudy Rucker also discusses the lower dimensions, makes analogies with the fourth dimension, and discusses *Flatland* in his popular introduction to the fourth dimension [A3].

1 Common Objections

Before embarking on our intellectual journey into the higher dimensions, let us pause briefly to consider the value of doing so. We will address some important points, such as time's role as an extra dimension and how there can be extra dimensions in our universe when it seems evident that we can only move in three independent directions. We will also see that higher-dimensional studies are not merely philosophical ventures. For one, the mathematical relationships involved have practical applications. For another, there are experimental tests underway with the potential to discover the presence of extra dimensions in our universe. Some of the concepts presented here will be explored more fully in later chapters, but should help to satiate your curiosity in the meantime.

1.0 Is Time the Fourth Dimension?

One objection to searching for a fourth dimension is that it has already been found: Time is the fourth dimension.

It is true that the dimensions of space and time are woven together to form spacetime in Einstein's special and general theories of relativity, and that time serves as a fourth dimension. However, time is a known dimension, like the first three dimensions of space; time is not an extra dimension.

A fourth spatial dimension or a second temporal dimension would be extra dimensions – i.e. beyond what has been observed in our universe. When referring to extra dimensions, they are generally assumed to be spatial since it is much less common to speak of extra dimensions of time.

Although time is a dimension, it is usually referred to explicitly as time. If you are setting a date for dinner, you might ask, "When would you like to meet?" Your question would probably be misinterpreted if instead you asked, "Where in the fourth dimension should we get together?"

Assume any extra dimensions to be spatial unless stated otherwise.

It might seem that three dimensions of space plus one dimension of time are adequate to describe our universe, but superstring theory, for example, provides compelling reasons to search for extra dimensions – i.e. more than three dimensions of space and one of time. Thus, time may very well not be the final dimension.

1.1 Where Are the Extra Dimensions?

Common experience shows that it is not possible to move into a fourth dimension. How can this be reconciled with any theory that proposes the existence of extra dimensions in our universe?

The nature of the extra dimensions may be different from the three known dimensions. This difference can explain how extra dimensions may not affect common experience.

There are two simple ways to have extra dimensions, different from the usual three dimensions, which would not have a noticeable effect on common experience. One way is that one or more of the particles observed in the known universe may not be able to propagate into the extra dimensions. This would explain why it is not possible to move into the extra dimensions. Secondly, the extra dimensions may be compact. If they are very tiny, this would explain why macroscopic beings do not notice them.

The motivation for these concepts originates from string theory. Since string theory is a theory of gravity – i.e. one goal is to unify gravity with the other three fundamental forces in a grand theory of everything – it is generally assumed that the mediators of the gravitational interaction, called gravitons, can propagate into all of the spacetime dimensions, including the extra dimensions. It is possible to develop models where some or all of the other particles are excluded from propagating into one or more extra dimensions.

For example, it could be that all of the known elementary particles are confined to what is termed the Standard Model wall. All of the particles observed in the known universe might be constrained to lie on a $(3+1)$-D wall, called a D_3-brane, except for the gravitons, which can propagate into extra dimensions. In this case, it is simply impossible for particles to propagate into the fourth and higher dimensions, explaining why the known universe appears to be 3D.

However, there are also models where some or all of the particles can propagate into one or more extra dimensions. In this case, the extra dimensions may be "hidden" through compactification. It is possible that in the early stages of the universe, shortly after the Big Bang, while three of the spatial dimensions grew to astronomical size, others were compactified; the extra dimensions may be compact, like a circle or ellipse, with a very small size. The idea of compactification can be exemplified simply with a long, thin cylinder (i.e. very long length, but very small radius): The length of the cylinder represents one of the usual dimensions, while the circumference represents an extra compact dimension.

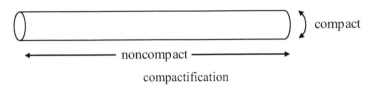

compact

←————————— noncompact —————————→

compactification

If the known universe has one extra compact dimension shaped like a circle, then a particle at any point in the known universe could travel through the extra dimension around this circle. If the size of this extra dimension is very small – smaller than the size of an atom, for example – then we would not be able to notice any particles moving through this circle, and the extra dimension would be hidden from our sight.

1.2 Why Study Extra Dimensions?

A common myth is that studies of extra dimensions can offer no practical value.

It might seem that higher-dimensional geometry might not be practical since macroscopic beings in the known universe are evidently unable to move in higher dimensions; and it might seem that the existence of small, extra, compact dimensions might be very important to superstring theory, but not practical otherwise. However, some of the mathematics of higher-dimensional geometry is actually used regularly in mathematics, science, economics, etc.

First, we will consider some practical applications of higher-dimensional mathematics. Following this, we will shift our attention to physics – namely, superstring theory.

A set of variables often forms a higher-dimensional relationship. If one variable w depends on three other independent variables x, y, and z, then the four variables w, x, y, and z form a 4D relationship. For example, w could be the cost of gasoline for a road trip, which depends on the cost per gallon of fuel x, the number of miles driven y, and the fuel efficiency of the car z in miles per gallon. In this case, $w = \dfrac{xy}{z}$. A graph of w as a function of x, y, and z shows the relationship between these variables. This is a 4D plot.

Such data can also be analyzed mathematically without even visually examining the full higher-dimensional plot. In fact, such data analysis is so routine that the analyst may not even think of it is as a higher-dimensional relationship.

A relationship between a set of variables forms an N-dimensional space. The variables may have absolutely nothing to do with the usual meaning of *space*. Many branches of science, economics, etc., deal with such vector spaces. The underlying mathematics – vector algebra, linear algebra, and tensor algebra – involves N-dimensional relationships. For example, the formula for finding the magnitude of a 4D vector is found using a formula similar to that used to find the length of the long diagonal of a tesseract, which involves the 4D generalization of the Pythagorean theorem. The mathematics of these higher-dimensional spaces is the same as the higher-dimensional geometry of what we usually mean by *space*.

Group theory is useful in many branches, of mathematics, physics, etc. For example, it is particularly useful in the study of crystals and theories of elementary particle physics. The same mathematics that applies to rotation groups for crystals and groups of elementary particles also applies to groups of corners of higher-dimensional polytopes.

On the physical side, superstring theory actually *predicts* the existence of extra dimensions. At first, this prediction was perceived as an objection to string theory, but the resolutions for how there can be extra dimensions in the known universe are becoming more accepted, in part because superstring theory itself has become more promising over the course of its development. Over the past ten years there has been a sudden explosion of professional research on superstring-inspired extra dimensions in the field of particle physics. Superstring theory is perceived by most as the best candidate presently available for a theory of everything. Including this, there are many aesthetic reasons for studying superstring theory and the associated extra dimensions.

This is an exciting time for superstring theory: One reason for the sudden increase in research activity in this area is that there are now reasonable prospects for experimental confirmation of the theory. Even many who are not engaged in the research on superstring-inspired extra dimensions are very interested in the results of current and upcoming experiments.

Aside from mathematical and physical reasons, contemplating the higher dimensions can result in a better understanding of 3D space and can sharpen and improve visualization skills.

Although there may be some practical value in studying the higher dimensions, interest in this field often does not stem from its potential usefulness. For many, the concepts associated with extra dimensions are a hobby, studied even though the research may not be viewed as practical. There are other compelling reasons to learn about extra dimensions, not the least of which are that extra dimensions may be interesting and fun to contemplate.

Finally, there are philosophical aspects and spiritual applications of extra dimensions. We will focus our attention on the mathematical and scientific properties, however.

1.3 How Can Extra Dimensions Be Detected?

Let us consider one final myth – that the extra dimensions of superstring theory are so small that they can never be observed.

The extra dimensions predicted by superstring theory were originally expected to have a size on the order of the Planck length, which is about 10^{-33} cm. However, recent developments have motivated prospects for large extra dimensions.

One or more of the extra dimensions predicted by superstring theory may be as large as a fraction of a millimeter; this may still seem small, but it is huge in comparison to the Planck length. Large extra dimensions are large enough for potential discovery in near-future experiments. It was not simply shown that the extra dimensions can be large: It is actually *motivated* as a possible solution to a fundamental hierarchy problem of supersymmetry. Just how large the extra dimensions can be depends on the details of a given model; a sub-millimeter is the extreme upper limit.

The potential for discovery and bounds on the size of large extra dimensions stem largely from astrophysical and cosmological observations, high-energy collider data, and experimental tests of Newton's law of universal gravitation. Since there is recent motivation for superstring-inspired extra dimensions to be much larger than originally predicted, current and upcoming physics experiments are underway that may detect the presence of extra dimensions in our universe.

Times have changed: The prospects for discovering extra dimensions and testing superstring theory are now much more plausible. Superstring theory is becoming more accepted and respected as a physically testable theory, rather than a philosophical consideration. Interest in superstring-inspired large extra dimensions is growing, as evidenced by the hundreds of professional research papers on their phenomenological ramifications that have been published in leading physics journals in the past decade.

Don't worry if some of these ideas seem Greek at this point. We will develop them more fully as we encounter them in later chapters.

Further Reading:

The original research papers that motivated large extra dimensions include work by Antoniadis [T1] and Arkani-Hamed, Dimopoulos, and Dvali [T2]. These papers are highly technical; for a much more accessible article on this topic, see [A4].

2 Visualizing the Fourth Dimension

We will begin our investigation of extra dimensions by considering a single Euclidean dimension – i.e. just like the three known dimensions. This will serve to establish methods for visualizing extra dimensions and comprehending the significance of the extra degree of freedom. It is also a necessary prelude to understanding compactification and other features of superstring theory and experimental searches.

2.0 The Fourth Dimension

A particle that can travel in 4D space can move along any linear combination of four independent directions: In addition to moving north/south, east/west, and up/down, there is a fourth direction in 4D space perpendicular to each of these other directions. This fourth independent direction is termed *ana*, and its opposite is called *kata*.

The three independent directions of 3D space can easily be visualized by examining the corner of a room: The three edges meeting at a corner are mutually perpendicular. These edges can serve as axes of a 3D coordinate system – one edge for x, one for y, and one for z.

Any other direction in 3D space is a linear combination of x, y, and z; any point in the room has a unique set of three independent coordinates (x, y, z). In 3D, it is not possible to orient a meterstick such that it is perpendicular to all three edges meeting at the corner: If the meterstick is perpendicular to any two edges, it will be parallel to the third.

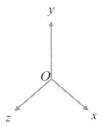

three edges meet at the corner of a 3D room

However, if the meterstick could move in the fourth dimension, then it would be possible to orient a meterstick such that it would be perpendicular to all three edges meeting at one corner of a 3D room. The meterstick would be parallel to ana. In fact, if it were possible to walk in the ana (or kata) direction, it would be very easy to exit the room – even if there were no windows or doors – without damaging the walls.

This can be seen by analogy: Imagine a 2D being in a 2D world trapped inside of a square. The being cannot escape the square by moving along any direction in the 2D space. However, if the being could move in the third dimension – perpendicular to the square – it would be very easy to escape. It would be very silly, though, to try to imprison a 3D being inside a square; it takes a 3D cell, perhaps in the shape of a cube, to trap a 3D being. It would take a 4D cell, such as a 4D cube – called a *tesseract* or a 4D *hypercube* – to imprison a being that could travel four-dimensionally.

A hypercube is an N-dimensional generalization of the cube. A tesseract is a special case of the hypercube corresponding to $N = 4$.

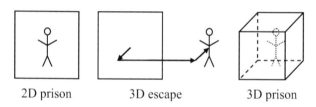

2D prison   3D escape   3D prison

It is challenging for 3D beings to fully understand 4D space; it is not easy to imagine moving in a direction that the known universe does not appear to permit. To see the problems with trying to grasp the extra freedom of a higher-dimensional space, try to imagine a 2D being contemplating what it would be like to live in a 3D universe.

2D being contemplating 3D space

While there is some ambiguity in interpreting a 2D drawing of a 3D object, there is additional ambiguity in representing a 4D object in 3D space. Drawing a 4D object on a 2D sheet of paper, for example, adds another degree of ambiguity compared to drawing 3D objects; the two independent directions of the sheet of paper are being utilized to represent four independent directions. In the case of drawing 3D objects, everyday experience helps to resolve this ambiguity. However, this is not the case for 4D objects. This makes it all the more challenging to attain a full understanding of 4D concepts.

However, through thought and effort, along with a guide (such as this book), it is possible to overcome these inherent challenges. Extra dimensions *can* be visualized and understood well. Conceptually, studies of the lower dimensions help to develop a better intuition for the higher dimensions. Techniques for visualizing extra dimensions, such as drawing or sculpting representations of higher-dimensional objects, can be developed through such analogies. The least ambiguous guide to the higher dimensions stems from mathematics. You may find yourself wondering, for example, "How can you be sure how many edges a tesseract has?" or, "How do you know that you can rotate your right hand in 4D such that it becomes a left hand?" The answer is generally seen by applying mathematics.

The most straightforward way to investigate higher dimensions is to generalize an algebraic equation to higher dimensions. For example, consider a square with length L and width W. The length of its long diagonal is found from the Pythagorean theorem to be $\sqrt{L^2 + W^2}$, and its area is LW. It is straightforward to show that a cube with length L, width W, and depth D has a long diagonal of $\sqrt{L^2 + W^2 + D^2}$ and a volume of LWD. The pattern continues quite naturally into the fourth dimension. A tesseract with length L, width W, depth D, and hyperdepth H has a long diagonal of $\sqrt{L^2 + W^2 + D^2 + H^2}$ and a hypervolume of $LWDH$. As a second example, in order to draw a 4D object, the natural starting place is a mathematical equation for the object, from which a 4D plot can be constructed and finally projected onto a lower-dimensional space.

One goal of this book is to present the concepts of extra dimensions clearly to all readers, assuming a minimal knowledge of mathematics (familiarity with the introductory elements of algebra and some basic geometric structures will suffice), while still engaging readers who have a stronger mathematical background. To this end, mathematics has been included where it is thought either to enhance understanding of the concepts or to be fascinating to the more mathematically minded. However, where there is math, it is usually supplementing a discussion that can engage readers who wish to skip the math. In cases where the math is a little more advanced, it is either explained for those who may not be familiar with it or it is presented more as an aside. There are a few entire sections devoted to mathematics – for those who are interested in such diversions – but these can reasonably be skipped with a minimal sacrifice to the content. Now we will examine various techniques for visualizing and understanding extra dimensions.

2.1 Projective Geometry

One way for 3D beings to visualize the fourth dimension is through projective geometry, which is like casting a shadow of a 4D object onto 3D space. Similar projections are made to draw 3D objects on a 2D sheet of paper. We see 2D projections when we view objects in the known universe, and we use 2D illustrations to represent what 3D objects look like. Our experience living in the physical universe helps us to interpret such 2D projections readily. Moreover, we are familiar with visualizing objects through projective geometry – from drawings, paintings, photographs, even vision itself – so we can apply our experience toward visualizing higher-dimensional objects.

Let us begin by considering projections for the three known dimensions. Imagine a being viewing objects in the second dimension. A 2D being would see 1D images of 2D objects. These 1D images are projections of 2D objects onto the 1D "surface" of an eye. These images could be drawn on a 1D canvas – namely, a line or an arc.

We can describe such a projection mathematically. A 2D world can be described by two independent coordinates, x and y. Just one axis alone, say x, is 1D. A 2D object can be projected onto the x-axis by dropping all of its points down onto the x-axis – i.e. each point of the 2D object is moved along the y direction (or $-y$ direction, if necessary) until it reaches the x-axis. This projection gives an idea of what a 2D eye would see if it were placed on the x-axis and looked in the y direction (or $-y$ direction, if the object is below the x-axis). The "surface" of the 2D eye is 1D, like the x-axis. Alternatively, a 2D object could be projected onto the y-axis, or onto any other line in the 2D space. This is easier to understand by studying the following diagrams.

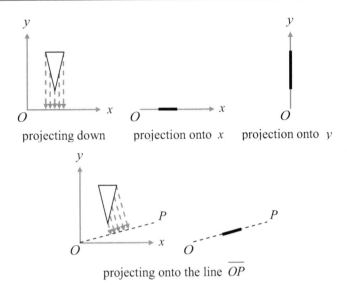

projecting down   projection onto x projection onto y

projecting onto the line $\overline{OP}$

In addition to the projection, depth perception, shadows, and color may help to distinguish between objects that would otherwise have identical projections.

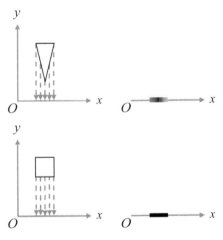

distinguishing between similar projections

The surface of a 3D eye is 2D. A 3D being sees 2D images, like a painting. A 3D world has three independent coordinates, x, y, and z. Excluding any one of these coordinates results in a plane, which is 2D. A 3D object can be projected onto the xy plane, the yz plane, the zx plane, or any linear combination of these.

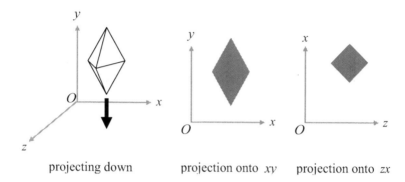

projecting down   projection onto xy projection onto zx

Multiple projections along different directions can combine together to provide more information about a higher-dimensional object. For a 3D object, for example, it is useful to project its image onto the xy, yz, and zx planes.

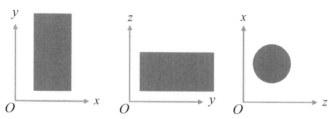

Puzzle 2.1: What 3D object would make these projections?

In the known universe, we view objects directly, but since there are evidently no 4D objects for us to view, their projections are more indirect – more like a shadow. If you illuminate a ball with a light source and view its shadow on a flat surface, its shadow will be a circle or an ellipse, depending upon the orientation of the surface relative to the incident light. Visualizing projections of 4D objects is analogous to a 2D being studying the shadow of a ball – keeping in mind that a 2D being does not see the circular shadow from same perspective from which we are accustomed to looking at drawn circles (rather, they would see it from an "edge" view). Nonetheless, these shadow-like projections still serve as useful visual aids.

45

There are two useful ways we might choose to study the projected images of a 4D object: We can project it onto a 2D sheet, or construct a 3D image. Drawing a 4D object on a 2D sheet of paper requires projecting the object onto a plane. A 4D world has four independent coordinates, x, y, z, and w. Choosing just two of these coordinates yields one of six planes: xy, zx, xw, yz, zw, or wy. Alternatively, the image of a 4D object can also be projected onto 3D space. There are four ways to choose three of the four independent coordinates: xyz, wxy, zwx, or yzw. These 3D images can be constructed in 3D space – e.g. using pipe cleaners or Play-Doh.

Let us consider the projections of two fundamental 4D objects – the glome and the tesseract. The glome (aka a hypersphere in 4D) is a generalization of the sphere to 4D space: The glome is the locus of points equidistant from a point (i.e. the center) in 4D space. The tesseract (aka a hypercube with $N = 4$) is a 4D generalization of the cube. Thinking about a tesseract is literally thinking outside of the box. We will analyze these objects in much more detail in subsequent chapters, but in this chapter we will use their essential features for illustrative purposes.

A hypersphere in N-dimensional space is the locus of points equidistant from a common center in N-dimensional space. The glome is a hypersphere in 4D space.

The circle, sphere, glome, hypersphere in 5D, and so on form a geometric pattern. Since the projection of a circle onto a line is a line segment (equal to the diameter of the circle) and the projection of a sphere onto a plane is a circle, we can reason that the projection of a glome onto a 3D hyperplane is a sphere. (This, along with many other qualitative observations that we will make, can be proven formally using mathematical techniques.)

The projection of the glome onto a plane is a circle, and the set of projections of the glome onto multiple planes helps to distinguish it from the projection of other objects. There is a similar problem in 3D: The projection of a finite right-circular cylinder onto a plane could be a circle or a rectangle, for example, so a couple of projections in combination help to distinguish the projection of the cylinder from the projection of a sphere or rectangular box.

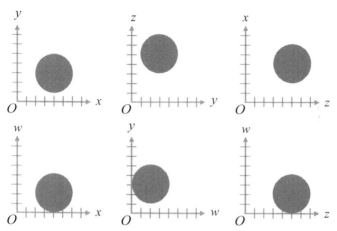

Puzzle 2.2: What are the coordinates of the center of this glome?

The tesseract is part of the following geometric pattern: square, cube, tesseract, 5D hypercube, etc. The projection of the square onto a line is a line segment, which could be as short as a line segment (where it is implied that the line lies in the plane of the square, since we have in mind a 1D being trying to visualize a 2D object in this case) or as long as the long diagonal of the square. In this case, we see that looking at the projection of a sphere or square onto two mutually perpendicular axes can result in two line segments of equal length, so a projection onto a third axis is needed to see the difference from a 1D perspective.

The projection of a cube onto a plane is a square if the plane is parallel to one of the sides, but otherwise it could also be a hexagon or irregular pentagon. (Don't confuse the use of the term *projection* that we are studying presently with *cross section*, which we will investigate later; we will find that there are more possible cross sections than the types of projections that we consider here.) Without yet defining a tesseract more precisely, we can reasonably expect the projection of a tesseract onto a 3D hyperplane to be a cube if the 3D hyperplane is parallel to one of the "3D sides" of the tesseract, but the projection may have more sides than a cube otherwise. We can also expect the projection of a tesseract onto a plane to be a square or polygon with more than four sides, depending upon the orientation of the tesseract relative to the plane.

These projections of the glome and tesseract offer mere glimpses of 4D objects, but we can improve upon this to visualize them better. For example, consider that when we draw a cube, we draw

47

more than the projection that we see. We add lines to represent the front edges, and dashed lines for the rear edges. This extra definition helps to resolve ambiguities. We are really drawing the third dimension by adding a diagonal axis to represent the depth. We can similarly draw the fourth dimension by adding two such diagonal axes.

2.2 Drawing the Fourth Dimension

When we draw a 2D object, we work with two independent coordinates, x and y. We generally draw the x-axis as horizontal and the y-axis as vertical. The two axes intersect at a point, which refer to as the origin. In order to add a third dimension, we need a third independent coordinate, z. We draw the z-axis diagonally (e.g. downward to the left) to represent the direction coming out of the page (so a negative value of z represents the direction going into the page). In this 3D coordinate system, there are three mutually orthogonal (perpendicular) axes, which all intersect at the origin. There are also three mutually orthogonal planes: the xy, yz, and zx planes. The intersection of any two of these planes is one of the coordinate axes. In order to draw the fourth dimension, we need one more coordinate axis – call it w, to represent the ana/kata direction.

On a 2D sheet of paper, if x is horizontal and y is vertical, then the other two directions, z and w, must both be perpendicular to the sheet of paper and perpendicular to each other. Of course, it is not possible in 3D space to find two lines perpendicular to a plane and to each other. However, it is possible in 4D space: z can come out of the page as usual, while w can go along the ana direction.

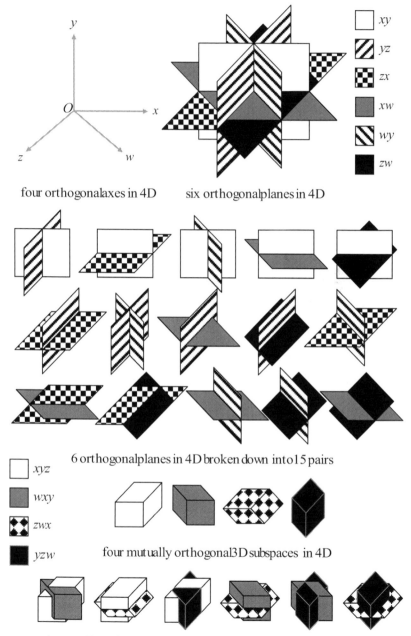

four orthogonal axes in 4D

six orthogonal planes in 4D

xy

yz

zx

xw

wy

zw

6 orthogonal planes in 4D broken down into 15 pairs

xyz

wxy

zwx

yzw

four mutually orthogonal 3D subspaces in 4D

4 mutually orthogonal 3D hyperplanes in 4D broken down into 6 pairs; in each case the region of intersection is a plane

The x-, y-, z-, and w-axes of a 4D graph are mutually orthogonal; the xy, zx, xw, yz, zw, and wy planes are similarly mutually orthogonal; and the 3D hyperplanes (3D hyperplanes of the 4D space) xyz, wxy, zwx, and yzw are also mutually orthogonal. Whereas the intersection of two orthogonal planes is a line, the intersection of two orthogonal hyperplanes is a plane.

A hyperplane is a higher-dimensional Euclidean space. The three known dimensions of our universe appear to form a 3D hyperplane. The term hyperplane is useful for designating a subspace. For example, the 3D *hyperplane* is an infinitesimal subspace of 4D space, just as a plane is an infinitesimal slice of 3D space.

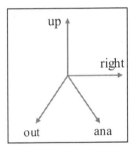

two lines (out and ana) perpendicular
to the plane of the paper and each other

In order to draw the four independent coordinates of 4D space on a 2D sheet of paper, at least two of the axes must be diagonal. For example, x may point to the right, with y up, z downward to the left, and w downward to the right. There is more ambiguity graphing in 4D compared to 3D because there are two degrees of freedom lost instead of just one. Diagonally down to the left, along the z-axis, represents out of the page as in a usual 3D graph, while diagonally down to the right, along the w-axis, represents the ana direction, unique to 4D space. 4D objects can be drawn on a 2D sheet of paper by plotting its defining points on a 4D graph, where each point has coordinates (x, y, z, w).

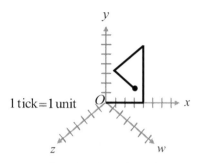

1 tick = 1 unit

locating the point (4,6,4,3)

A tesseract is a 4D generalization of the 3D cube. The coordinates of the corners of the unit tesseract are binary – i.e. x, y, z, and w can each be zero or one. Thus, a tesseract has 16 corners. The corners are connected by edges that run parallel to the x-, y-, z-, or w-axes.

(0,0,0,0)	(0,1,0,0)	(1,0,0,0)	(1,1,0,0)
(0,0,0,1)	(0,1,0,1)	(1,0,0,1)	(1,1,0,1)
(0,0,1,0)	(0,1,1,0)	(1,0,1,0)	(1,1,1,0)
(0,0,1,1)	(0,1,1,1)	(1,0,1,1)	(1,1,1,1)

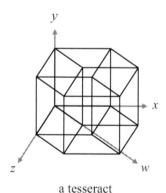

a tesseract

To see that the unit tesseract is indeed a geometric extension of the unit cube, try this: Delete the fourth entry from each corner's coordinates and you will obtain 8 distinct points – the coordinates of a cube. If instead you delete the second coordinate (or the first or third, so long as you are consistent), you will obtain a cube in a different 3D hyperplane. By deleting two coordinates – e.g. the third and fourth – you will obtain 4 distinct points, corresponding to a square. On the other hand, if you had a fifth coordinate, you would obtain the 32 corners of a 5D hypercube.

Here is one prescription for drawing a tesseract:

1. First draw a square in the xy plane.
2. Draw a replica of this square shifted diagonally down to the left (along the z direction). The upper right corner of the new square should not touch an edge of the first square.
3. Draw 4 lines diagonally down to the left to form a 3D cube.
4. From each of the 8 corners of this cube, draw a line diagonally down to the right (along the w direction). None of these 8 lines should end on any of the previous edges.
5. Add 4 horizontal, 4 vertical, and 4 " z " lines.

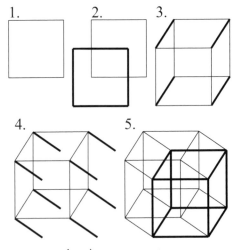

drawing a tesseract

2.3 Constructing the Fourth Dimension

Compared to drawing 2D images of 4D objects, it is a significant advantage to construct 3D images. This can be achieved via sculpture, woodwork, metalwork, etc. Some simple, clean, and cost effective suggestions include: tying pipe cleaners together, joining toothpicks or long thin metal rods together by poking them into small Styrofoam balls, gluing wooden dowels together, and shaping Play-Doh. Using 3D space to create a graph, x can point to the right, y can point forward, and z can point upward. While x and y may lie on the surface of a table, the problem is suspending a point above the table to represent the z coordinate. This is why it is necessary to use materials that allow for vertical supports. In such a spatial graph, x, y, and z are all perpendicular, avoiding some of the ambiguities in drawing on a sheet of paper. However, a diagonal direction is still required for the fourth dimension: w could point diagonally left, upward, and backward, for example. Since w will not be perpendicular to either x or y in such a 3D representation, an image formed by plotting the defining points of a 4D object will be a projection onto 3D space. This is analogous to using a 2D sheet of paper to draw the third dimension.

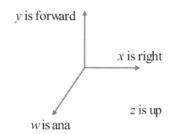

top view of a 4D graph in 3D space

It is useful to combine projections together to depict a higher-dimensional object. For example, consider a 3D cube with edges parallel to x, y, and z. If this object is projected onto the xy, yz, or zx plane, the projected image will be a square. Doubling these projections yields the six squares that bound the cube.

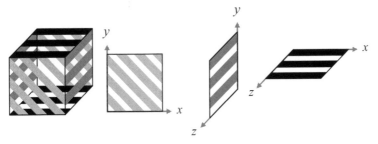

a 3D cube and its projections onto the *xy*, *yz*, and *zx* planes

Similarly, for a tesseract with its edges parallel to x, y, z, and w, its projections onto the 3D hyperplane *xyz*, *wxy*, *zwx*, or *yzw* are cubes. Doubling these projections yields the eight cubes that bound the tesseract.

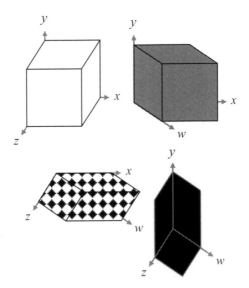

the projections of a tesseract onto the
xyz, *wxy*, *zwx*, and *yzw* 3D hyperplanes

These 3D projections of a tesseract can be constructed in 3D using pipe cleaners for the edges, for example. The *xyz* projection is the only one that will look like a cube, with right angles and edges of equal length. The projections involving the w-axis will have some acute or obtuse angles. In 4D space, these would all be perfect cubes. Constructing these 3D projections can aid in trying to understand the tesseract.

A double set of these projections can be put together to form a tesseract, but keep in mind that the eight cubes that bound the tesseract share several edges. A simpler way to construct a 3D representation of a tesseract is to make two interlocking cubes, then connect the corners with appropriate edges.

constructing a 3D representation of a tesseract

2.4 Higher-Dimensional Perspective

Imagine looking down a long hallway. The edges of the hallway are physically parallel, but in the 2D image seen through eyes, the edges do not appear to be parallel. Similarly, when viewing a cube, the front face appears larger than the rear face; hence, the connecting edges are not all parallel. In order to draw a 2D representation of a 3D object the way it appears when viewed through eyes, it is necessary to account for perspective. There is a similar problem in using 2D or 3D to represent higher-dimensional objects.

looking down a hallway   front face appears larger

Consider a square lying in the xy plane with its edges parallel to the x- and y-axes. A projection onto the x-axis results in a line segment. Every point on the lower edge of the square projects the same distance vertically down to the x-axis. Projecting onto an eye is not quite the same as projecting onto an axis. The eye is small compared to most of the objects it views. If a 2D being viewed the square from the x-axis, and if the lower edge were partially transparent, the 2D being would perceive the upper edge to be shorter.

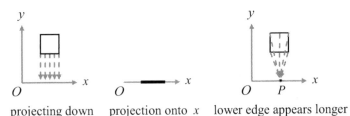

projecting down   projection onto x lower edge appears longer

In order to draw 2D representations of 3D objects using perspective, lines of depth converge to a single point – the vanishing point P. The lines of depth depend on the location of P, which is related to the perspective from which the object is viewed. Assuming that the object is viewed from the front, if P lies to the left of the object, the object is viewed from the left of the object; if P lies above the object, the object is viewed from above; and so on.

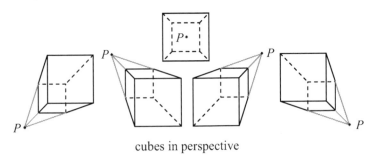

cubes in perspective

In order to show perspective in a 3D representation of a 4D object, the lines of "hyperdepth" – i.e. the diagonal lines representing the ana/kata direction – should converge at a vanishing point, even though these ana/kata lines are parallel for the actual object in 4D space.

When showing perspective in a 2D representation of a 4D object, two vanishing points are needed – one for the diagonal z direction, for the depth into/out of the page, and another for the diagonal w direction, for the hyperdepth along ana/kata.

 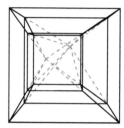

tesseracts in perspective

In general, $(N - M)$ vanishing points are needed to represent an N-dimensional object on a M-dimensional subspace. For example, drawing a 5D object with a 2D sheet of paper requires $5 - 2 = 3$ vanishing points.

2.5 Lower-Dimensional Analogies

Knowledge and experience with the lower dimensions can provide much insight into the higher dimensions. For example, there are many patterns going from 0D to 1D to 2D to 3D, which help to make predictions in 4D, 5D, and so on.

The following pattern leads to the concept of a tesseract: point, line segment, square, cube, tesseract, 5D hypercube, etc. Consider some similarities in how these objects are formed: In 1D, two points are joined by a line segment; in 2D, two opposite edges are joined by two line segments; and in 3D, two opposite squares are joined by four more squares. By analogy, in 4D, two opposite cubes are joined by six more cubes (so there are eight cubes in all), completing the following pattern: two joining line segments in 2D, four joining squares in 3D, six joining cubes in 4D. There is another pattern in the corners: 1 corner in 0D, 2 in 1D, 4 in 2D, and 8 in 3D, which leads to 16 corners in 4D.

Some patterns are more subtle. For example, count the edges: 0 in 0D, 1 in 1D, 4 in 2D, and 12 in 3D. From this short pattern alone, it is not so obvious how many edges a tesseract has. It turns out that a tesseract has 32 edges. Here is how it works: In 1D, there is 1 edge along the x-axis; in 2D, there are 2 edges along x plus 2 more along y; in 3D, there are 4 edges along x, 4 along y, and 4 along z; and in 4D, there are 8 edges along each of 4 axes, yielding 32 edges in all. Hypercube patterns are discussed in detail in Chapter 3.

There are also geometric patterns. For example, drawing a 3D cube on a 2D sheet of paper, its outline is generally a hexagon. Drawing a tesseract on a 2D sheet of paper, its outline is generally an octagon. The outline of a 5D hypercube is generally a decagon, and so on.

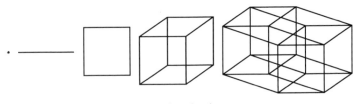

N-dimensional cubes

Defining the circle as the locus of points equidistant from a common point (the center of the circle) in 2D space, the circle itself is a 1D object (it has no interior). Modifying the definition to include all points in 3D space equidistant from the center generalizes the circle to a 2D sphere. In 4D space, it is called a *glome*. In N-dimensional space, it is termed a *hypersphere*. Walking along a circle, there is only one degree of freedom – clockwise or counterclockwise. Walking on the surface of a sphere, there are two degrees of freedom – changing latitude or longitude. Walking on the hypersurface of a glome, there would be three independent directions, which could be parameterized by north/south, east/west, and hypereast/hyperwest; the fourth direction, up/down, not being possible if confined to the hypersurface – just like up/down are not possible for an object confined to the surface of a sphere because up/down equates to getting closer to or further from the center of the sphere.

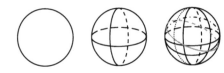

spheres in N-dimensional space

The sphere is not the only way to generalize a circle to 3D. For example, the cylinder is also based on the definition of a circle.

Instead of modifying the definition to include all points in 3D space equidistant from a common point, changing it to all points in 3D space equidistant from a common axis instead yields a right-circular cylinder (which can be finite by imposing an additional constraint on height, and adding ends). If the radius is proportional to the axial coordinate, a cone results. There are more possibilities in 4D. For example, the locus of points in 4D space equidistant from an axis results in a cylindrical type of object with spherical cross section; another possibility is the locus of points in 4D space equidistant from a plane, which has circular cross section.

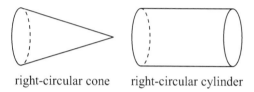

right-circular cone right-circular cylinder

2.6 Snapshots of the Fourth Dimension

Time is a concept associated with motion. If every particle in the universe were perfectly stationary, there would be no time. Time is a measure of change. Measurements of time are compared with patterns in nature – e.g. celestial motions, such as the rotation of the earth about its axis or the revolution of the earth around the sun; periodic motions, like an oscillating pendulum; or atomic patterns, such as radioactive carbon dating.

Time's role as a fourth dimension can be exploited to gain a better understanding of what a fourth spatial dimension may be like. The known universe has one time dimension and at least three spatial dimensions. These four dimensions are similar, in a way, to having four spatial dimensions, where time is ignored. With four spatial dimensions plus time, the extra dimension can be studied, for example, by looking at the effect that the motion of a 4D object has on 2D or 3D representations of the object. If the 4D object rotates, for example, this may cause changes in its lower-dimensional projections.

Including time in a plot requires one more dimension for the graph. For example, consider an object moving forward and/or backward in 1D. To graph its position as a function of time requires a 2D graph, just like plotting y as a function of x. To graph the position of an object winding through 3D space as a function of time requires a 4D plot.

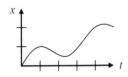

Another way to illustrate time is to collect a set of snapshots. Objects can be animated in flipbooks by drawing on each page the way the object looks at that moment. Flipping through the pages rapidly, the objects appear to move. A movie reel works in much the same way. Collecting a set of 2D or 3D snapshots of a 4D object as it moves through 4D space provides more information about the object than looking at a single image. A set of sculptures, for example, can serve as 3D snapshots of the fourth dimension.

A computer or television monitor can animate objects, adding a third, temporal dimension to the image shown on a 2D screen. Animation can similarly be utilized to show the motion of a projection of a 4D object on a 2D screen. It is also possible to animate in 3D, though perhaps not nearly as convenient as 2D.

Somewhat different from a projection, it can also be useful to examine cross section – the intersection of an object and a lower-dimensional subspace. For example, the intersection of a plane and a sphere is a circle (or a point); the size of the circle depends on where the plane slices the sphere.

cross sections of a sphere

A pure translation occurs when every part of an object moves in the same direction. Compare with rotation, where different parts of an object are moving in different directions with different speeds. Consider, for example, a sphere passing through a plane. When the sphere just touches the plane, the cross section is a single point. As the sphere passes through the plane, the cross section is a circle that grows until the sphere is halfway across, then diminishes back to a single point before vanishing. A glome passing through 3D space along the ana direction would behave, similarly, as an enlarging and then reducing sphere.

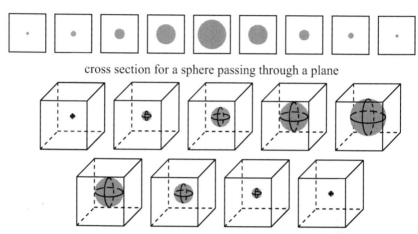

cross section for a sphere passing through a plane

cross section for a glome passing through 3D space

In rotation, each part of a rigid body travels in a circle about a particular axis. For a rigid body rotating in 2D space, the "axis" is a point. For rotations in 3D space, the axis is a line. In 4D space, rigid bodies rotate about a plane. In N-dimensional space, the "axis" of rotation is an $(N-2)$-dimensional subspace.

a triangle rotating in 2D space

Imagine a cube centered about the origin with its edges initially parallel to the x-, y-, and z-axes, and consider its cross section in the xy plane for potential rotations. Initially, the cross section is a square. If the cube rotates about the z-axis, the square cross section will also rotate. If the cube rotates about the x- or y-axis, the cross section does not rotate, but becomes an expanding rectangle until rotating through 45°, then contracts for 90°, expands for 90°, etc. There are other possible rotation axes – e.g. about a face

61

diagonal or body diagonal. Depending upon the rotation axis, the cross section of a rotating 3D cube can be a polygon with as many as six edges, and the number of edges of the cross section may change during the rotation.

In 4D, a rigid body rotates about a plane. This may seem strange, since in 3D rigid bodies rotate about an axis, which is a line. When a rigid body undergoes a simple rotation, each part of the rigid body travels in a circle. This circle defines a plane of rotation. A 2D space is already a plane, so every part of a rigid body in 2D travels around a point (located at the center of rotation). Imagine a rigid body in 3D rotating about the z-axis: Every part of the rigid body travels in a circle parallel to the xy plane, except for any points lying on the z-axis, which remain stationary on the z-axis. Even in 4D, every part of a rigid body travels in a circle. If every part of a 4D rigid body travels in a circle parallel to the xy plane, then no parts of the rigid body are moving in the z or w directions; since the z and w coordinates remain unchanged, the object is said to rotate about the zw plane. For a rotation about the yz plane, every part of the rigid body would travel in a circle in the wx plane.

Imagine a tesseract centered about the origin with its edges initially parallel to the x-, y-, z-, and w-axes, and consider its 3D cross section in the xyz hyperplane for potential rotations. The 3D cross section is obtained by setting w equal to zero, just as a 2D cross section of a cube in the xy plane is found by setting z to zero. Initially, the 3D cross section is a cube. For a rotation about the wx, yw, or zw plane, the cubic cross section rotates, while for rotations about the xy, yz, or zx plane, two sets of parallel sides expand and contract. For other "axes" (i.e. planes) of rotation, the 3D cross section can be a polyhedron with as many as 8 sides.

Further Reading:

For a classic text on projective geometry, see Coxeter's work [A5].

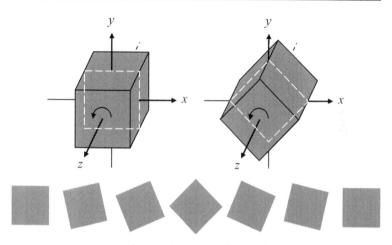

cube rotating about the z-axis

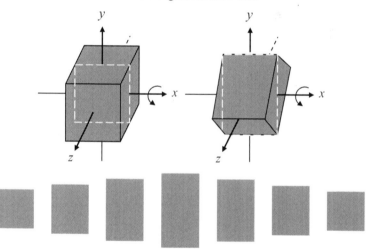

cube rotating about the x-axis

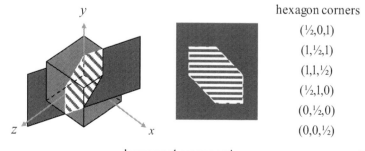

hexagon corners

($\frac{1}{2}$,0,1)

(1,$\frac{1}{2}$,1)

(1,1,$\frac{1}{2}$)

($\frac{1}{2}$,1,0)

(0,$\frac{1}{2}$,0)

(0,0,$\frac{1}{2}$)

hexagonal cross section

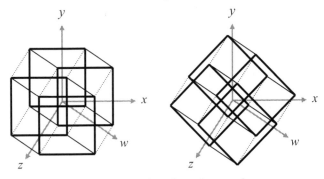

a tesseract rotating about the *wz*-plane

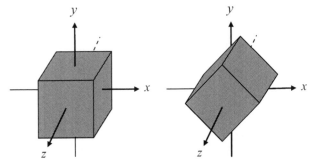

3D cross sections for a tesseract rotating about the *wz*-plane

3 Higher-Dimensional Polytopes

In this chapter, we will explore fundamental higher-dimensional geometric structures with flat sides – generalizations of polyhedra. In particular, we will study the hypercube in much detail since its right-angled nature provides much insight into the significance of the extra degrees of freedom available in higher dimensions. The hypercube also provides an idea of what basic engineering structures would be like if the known universe had a fourth dimension just like the other three. So, extra time spent analyzing the hypercube now will pay dividends later. We will also examine other basic higher-dimensional structures, since many objects in a higher-dimensional world would look much different from a hypercube. There are a couple of mathematical interludes in this chapter for those who enjoy patterns, which are easily skipped by those who wish to focus more on the concepts.

3.0 Polygons, Polyhedra, and Polychora

An N-dimensional closed shape with flat sides is called a polytope. The Schläfli symbol serves as a mathematical means of distinguishing between various polytopes – more informative than calling them by their names. The Schläfli symbol includes $N-2$ integers separated by commas between two braces; it has the form $\{p,q,r...\}$. We will learn how the various integers are determined as we encounter various polytopes.

   The simplest polytope is the polygon, which is a closed figure in 2D with $n \geq 3$ straight edges. The Schläfli symbol for the polygon has a single element of the form $\{p\}$, where p is the number of edges. Some common polygons include the triangle $\{3\}$, quadrilateral $\{4\}$, pentagon $\{5\}$, hexagon $\{6\}$, and octagon $\{8\}$. Regular polygons, such as the equilateral triangle and the square, are both equilateral and equiangular.

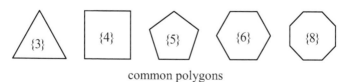

common polygons

65

A polyhedron is a closed surface in 3D with flat sides (faces), the $n \geq 4$ faces of which are polygons. For example, 4 equilateral triangles form a tetrahedron, 6 squares form a cube, and 12 pentagons form a dodecahedron. Regular polyhedra have identical faces arranged similarly at each vertex. The Schläfli symbol for a polyhedron has the form $\{p,q\}$, where p equals the number of edges of the polygonal faces and q equals the number of faces meeting at each vertex. For example, $\{3,3\}$ is a polyhedron constructed from triangles, where three triangles meet at each vertex; thus, $\{3,3\}$ is a tetrahedron. The cube is represented by $\{4,3\}$ and the dodecahedron by $\{5,3\}$.

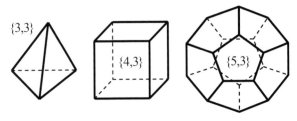

some regular polyhedra

A polychoron is a closed hypersurface in 4D bounded by $n \geq 5$ polyhedra. Examples include the pentachoron – a generalized tetrahedron – and the tesseract. For a polychoron, the Schläfli symbol has the form $\{p,q,r\}$, where p is the number of edges of each polygonal face, q equals the number of faces that meet at each vertex in the polyhedron cells, and r is the number of polyhedra meeting at each edge of the polychoron.

The polygon (2D), polyhedron (3D), and polychoron (4D) are special cases of the polytope. In general, the "sides" of an N-dimensional polytope are $(N-1)$-dimensional polytopes. For example, a polyhedron is 3D, so $N = 3$, and its sides are polygons, which are 2D. This checks out since $3 - 1 = 2$. It is important to note that we assume polytopes to be solid – i.e. include the interior region – when determining their dimensionality, in contrast to curved surfaces like spheres and cylinders. We say that a cube is 3D, while a sphere is 2D, because we take the sphere to be hollow, but the cube to be solid. In this regard, the cube is more like the ball, which is a solid sphere.

The Schläfli symbol has the form $\{p,q,r,...\}$, where p denotes the number of edges of the polygons from which the polytope is based, q indicates how many such polygonal faces meet at each vertex in the polyhedron cells, r represents the number of polyhedra meeting at each edge in the polychoron cells, etc. Add one to the number of elements in the Schläfli symbol to determine the dimensionality of the polytope.

For example, consider the Schläfli symbol for the octahedron, which is $\{3,4\}$. There are two elements in this symbol – a 3 and a 4. Since $2+1=3$, this is a 3D polytope – i.e. a polyhedron. The first element indicates that the sides of the polyhedron are triangular, and the second number indicates that 4 triangles meet at each vertex. The symbol $\{3,4\}$ is much more informative than the name, *octahedron*.

As a second example, consider the cube. Since the cube is a polyhedron, its Schläfli symbol must have two elements. Each side of a cube is a square, so the first element is a 4. Four squares meet at each vertex of a cube, so the second element is a 3. Thus, the Schläfli symbol for a cube is $\{4,3\}$. We will encounter more examples as we move along.

3.1 Anatomy of a Tesseract

The Schläfli symbol for a tesseract (aka 8-cell, octahedron, or 4D hypercube), $\{4,3,3\}$, defines the tesseract as a regular polychoron based on a regular 4-sided polygon (i.e. the square), in which 3 squares meet at each corner to form the regular polyhedron cell (namely, the cube) and 3 cubes meet at each edge of the tesseract. This can be checked by inspecting the following figure.

Four equal line segments connect at right angles to form a square. The cube generalizes the square to 3D: Six squares connect at right angles to form a cube. The tesseract is a 4D generalization of the cube. Eight cubes connect at right angles to form a tesseract.

The corners of the N-dimensional generalization of the unit cube can be defined as the set of points where x, y, z, w, etc. are binary – i.e. they are each either a 0 or a 1. The unit square has 4 corners: $(0,0)$, $(0,1)$, $(1,0)$, and $(1,1)$. The unit cube has 8 corners: $(0,0,0)$, $(0,0,1)$, $(0,1,0)$, $(0,1,1)$, $(1,0,0)$, $(1,0,1)$, $(1,1,0)$, and $(1,1,1)$. The unit tesseract has 16 corners: $(0,0,0,0)$, $(0,0,0,1)$, $(0,0,1,0)$, $(0,0,1,1)$, $(0,1,0,0)$, $(0,1,0,1)$, $(0,1,1,0)$, $(0,1,1,1)$, $(1,0,0,0)$, $(1,0,0,1)$, $(1,0,1,0)$, $(1,0,1,1)$, $(1,1,0,0)$, $(1,1,0,1)$, $(1,1,1,0)$, and $(1,1,1,1)$.

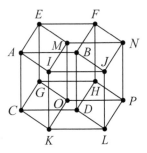

16 corners of a tesseract

A square has 4 edges – for a unit square with binary coordinates, two edges are parallel to the x-axis and two are parallel to the y-axis. A cube has 12 edges – four parallel to each of three axes. A tesseract has 32 edges – eight parallel to each of four axes:

$// x$		$// y$		$// z$		$// w$	
$\overline{BA}$	$\overline{JI}$	$\overline{CA}$	$\overline{KI}$	$\overline{AE}$	$\overline{IM}$	$\overline{AI}$	$\overline{EM}$
$\overline{DC}$	$\overline{LK}$	$\overline{DB}$	$\overline{LJ}$	$\overline{BF}$	$\overline{JN}$	$\overline{BJ}$	$\overline{FN}$
$\overline{FE}$	$\overline{NM}$	$\overline{GE}$	$\overline{OM}$	$\overline{CG}$	$\overline{KO}$	$\overline{CK}$	$\overline{GO}$
$\overline{HG}$	$\overline{PO}$	$\overline{HF}$	$\overline{PN}$	$\overline{DH}$	$\overline{LP}$	$\overline{DL}$	$\overline{HP}$

A cube has 6 square faces – for a unit cube with binary coordinates, two faces are parallel to the xy plane, two faces are parallel to the yz plane, and two faces are parallel to the zx plane. A tesseract contains 24 squares – four squares parallel to each of 6 planes; for the "binary unit tesseract," these are the xy, zx, xw, yz, zw, and wy planes:

$// xy$	$// yz$	$// zw$	$// xw$	$// zx$	$// wy$
$ABDCA$	$AEGCA$	$AEMIA$	$ABJIA$	$ABFEA$	$AIKCA$
$EFHGE$	$BFHDB$	$BFNJB$	$EFNME$	$CDHGC$	$BJLDB$
$IJLKI$	$IMOKI$	$CGOKC$	$CDLKC$	$IJNMI$	$EMOGE$
$MNPOM$	$JNPLJ$	$DHPLD$	$GHPOG$	$KLPOK$	$FNPHF$

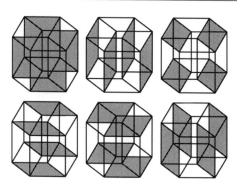

24 faces of a tesseract

Many of these faces may not look square since two of the axes are drawn at angles to depict the third and fourth dimensions. Bear in mind that x, y, z, and w are all mutually perpendicular.

A square is a 2D area bounded by 4 edges – two edges perpendicular to each of two axes. A cube is a 3D volume bounded by 6 sides – two faces orthogonal to each of three axes. A prisoner in a rectangular 4D prison would be "boxed in" from 8 "sides," where each "side" is a cube. A cube serves as a wall in 4D; compare with 2D, where a line segment can be a wall, and 3D, where a square can act as a wall. A tesseract is a 4D hypervolume bounded by 8 cubes – two cubes orthogonal to each of four axes:

$\perp x$	$\perp y$	$\perp z$	$\perp w$
AIKCEMOG	ABJIEFNM	ABJICDLK	ABDCEFHG
BJLDFNPH	CDLKGHPO	EFNMGHPO	IJLKMNPO

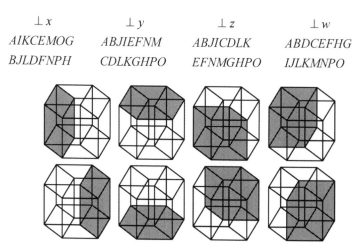

8 cubes of a tesseract

2 edges intersect at each corner of a square, 3 edges intersect at each corner of a cube, and 4 edges intersect at each corner of a tesseract. 2 squares intersect at every edge of a cube, and 3 squares intersect at every edge of a tesseract. 2 cubes intersect at every square of a tesseract.

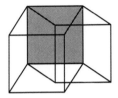

two orthogonal cubes intersect at a square (impossible in 3D)

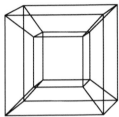

Puzzle 3.1: For this tesseract drawn in perspective:
- Find the eight cubes. Which cubes are parallel?
- Find the thirty-two edges. Which edges are parallel?
- Find the twenty-four squares. Which squares are parallel?
- Where are the two perspective points?

For a plane intersecting a cube, the cross section can be an infinitesimal point if it intersects at just one corner, a line segment if it intersections along just one edge, or a polygon with three to six sides, depending upon the orientation and position of the plane with respect to the cube. The intersection of a 3D hyperplane and a tesseract can be a point, a line segment, a square (if it intersects at just one face), or a polyhedron with four to eight faces (not necessarily of the same type of polygon – e.g. some of the faces may be triangles and others may be quadrilaterals).

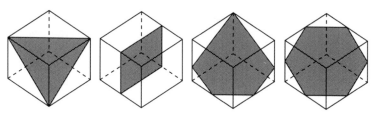

various 2D cross sections of cubes

Puzzle 3.2: Is it possible for the cross section of a cube to be a trapezoid? If so, how?

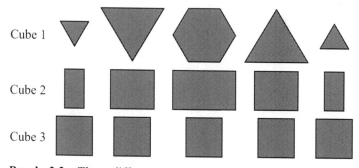

Puzzle 3.3: Three different cubes passing through a plane leave the cross sections illustrated above. Explain how these different cross sections are formed.

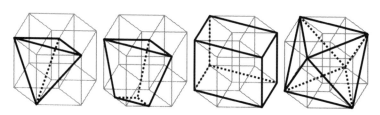

various 3D cross sections of a tesseract

Consider a polygon formed by the intersection of a plane with a cube: Each edge of the polygon lies in one of the six faces bounding the cube. This is why the polygon can have a maximum of six sides. The hexagon cross section has an edge in each of the six faces, whereas the triangular cross section has edges in only half of the faces. Now consider a polyhedron formed by the intersection of a 3D hyperplane with a tesseract: Each polygonal face of the polyhedron lies in one of the eight bounding cubes. In the case of the octahedron, the cross section includes a polygonal face in all eight bounding cubes; the tetrahedron uses just half the cubes.

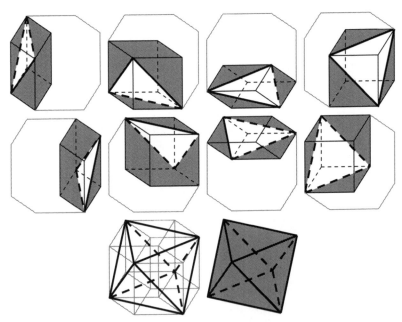

each side of a 3D cross section of a tesseract lies in one of the 8 bounding cubes

As a moving cube passes through a plane, its translation produces a series of 2D cross sections. Similarly, the translation of a tesseract can be depicted as a series of 3D cross sections.

When a cube rotates about an axis, every point of the cube travels in a circle around the axis – except for a point on the axis itself, which remains stationary. Such a circle defines a plane of rotation. Consider a cube centered about the origin initially with edges parallel to the coordinate axes. If the cube rotates about the z-axis, parts of the cube with the same z-coordinate travel in concentric circles within the same plane of rotation, while parts of the cube with different z-coordinates travel in circles in parallel planes of rotation. Each plane of rotation is perpendicular to the axis of rotation (the z-axis, for this example).

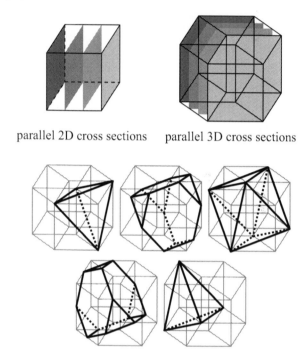

parallel 2D cross sections   parallel 3D cross sections

3D cross sections for a tesseract passing through a 3D hyperplane

Whereas a cube rotates about an axis, a tesseract rotates about a plane. Any point lying on this plane remains stationary, while every other point travels in a circle. The circle defines another plane – a plane of rotation – not to be confused with the stationary plane that serves to define the rotation (it is the generalization of the axis to a 4D rotation). For example, for a rotation about the zw plane, every point travels in a circle in the xy plane. Hence, the x- and y-coordinates of each point change, while the w- and z-coordinates remain unchanged. Also, the 16 edges lying in the xy plane rotate, while the 16 edges parallel to the w- and z-axes remain parallel to the

73

respective w- and z-axes. However, for a general plane of rotation, all 32 edges may change direction.

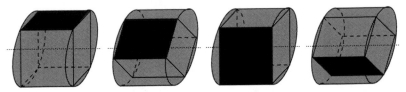

each point on a rotating cube travels in a circle perpendicular to the axis of rotation

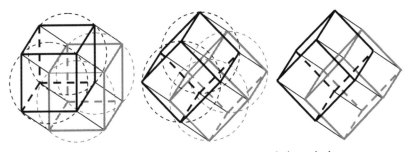

each point on a rotating tesseract travels in a circle

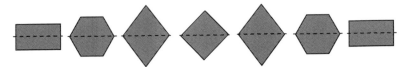

Puzzle 3.4: How can a cube rotate and yield the above 2D cross sections?

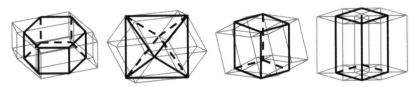

3D cross sections for a rotating tesseract

Consider the intersection between a plane and a cube, which represents a given cross section of the cube. As the cube rotates, the cross section will change size and/or shape. Since the rotation between the plane and cube is relative, there are two ways to think of it: From the plane's perspective, the plane is stationary and the cross section of the cube rotates; from the cube's perspective, the cube is stationary and the plane rotates. The relative rotation is the same apart from a directional minus sign – i.e. if the cube rotates clockwise relative to a stationary plane, then drawing the cube stationary, the plane rotates counterclockwise. There is an analogous equivalence between a rotating tesseract and an intersecting 3D hyperplane.

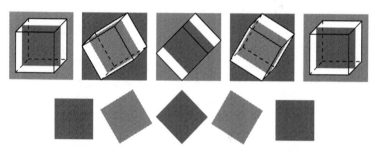

cross section relative to plane for stationary plane, rotating cube

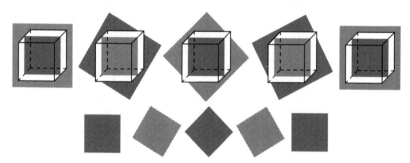

cross section relative to plane for stationary cube, rotating plane

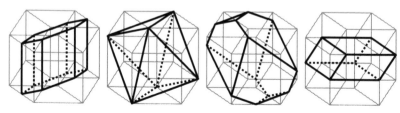

rotating 3D cross sections for a fixed tesseract

Another way to visualize a higher-dimensional object is to unfold it into a lower-dimensional space. For example, a cube can be unfolded into a 2D cross-like arrangement of 6 squares. Similarly, a tesseract can be unfolded into a 3D double cross-like arrangement of 8 cubes.

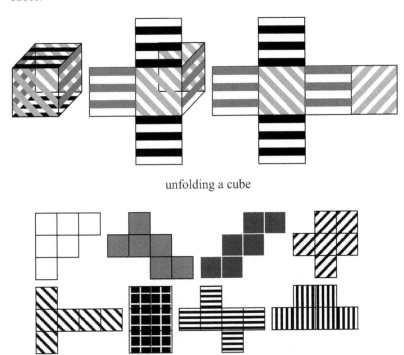

unfolding a cube

Puzzle 3.5: Which of these arrangements can be folded into a cube?

Lower-dimensional objects can be reflected upon rotation through a higher-dimensional space. For example, consider a 2D L-shaped object where one leg is longer than the other. This L-shaped object can merely rotate by moving within the 2D space; in order to reflect – to appear ⌐-shaped – it must move in the third dimension. By picking the "L" up off the page, turning it over, and replacing it on the page, it would look like an "⌐." Similarly, a 3D object can reflect upon a rotation in 4D space.

Such reflections are evident in the unfolded tesseract, where a cube can rotate in ways not possible in 3D space. Unfolding a cube, a 2D face hinges about an edge; when unfolding a tesseract, a cube hinges about one of its 2D faces (since in 4D objects rotate about a plane, not a linear axis). A square confined to a plane cannot rotate about an edge – this is only possible if the square can move into a third dimension. Similarly, a cube confined to 3D space cannot rotate while one face remains hinged – this is only possible if the cube can move in the ana/kata directions. Draw an object on a sheet of paper and turn it over in 3D, and the object will appear backwards – in a way that would generally not be possible by simply moving the object around within the plane of the sheet. Similarly, a 3D object turned over in 4D appears backwards – in such a way that the same reflection generally cannot be achieved in 3D space.

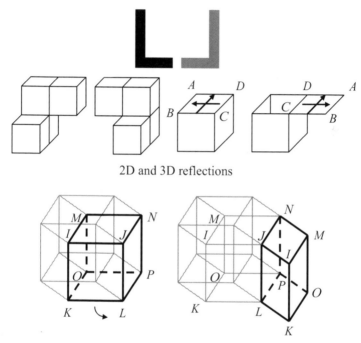

2D and 3D reflections

reflection of a cube rotating through 4D

Puzzle 3.6: Could this block reflect as shown upon a rotation in 4D space?

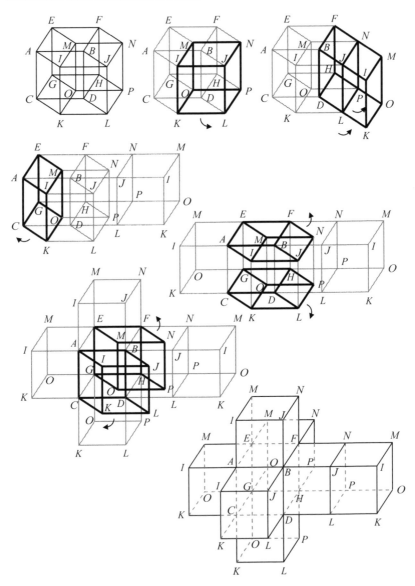

unfolding a tesseract

3.2 Hypercube Patterns

A hypercube is a higher-dimensional generalization of the cube. The special case of a 4D hypercube corresponds to the tesseract. Here is a prescription for how to draw a 5D hypercube:
1. First draw a tesseract.
2. Attach 16 congruent line segments to each corner of the tesseract. These lines should not be parallel to any of the edges of the tesseract, as they correspond to the fifth dimension.
3. Add four sets of parallel edges – corresponding to the four independent directions of the tesseract – to complete the 5D hypercube. An N-dimensional hypercube can similarly be drawn by starting with an $(N-1)$-dimensional hypercube.

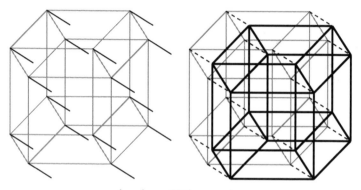

drawing a 5D hypercube

An N-dimensional cube has 2^N corners, since there are N coordinates (x, y, z, w, etc.) – each of which can be 0 or 1 for the N-dimensional binary unit cube. A point is its own corner since $2^0 = 1$, a line segment has $2^1 = 2$ endpoints, a square has $2^2 = 4$ corners, a cube has $2^3 = 8$ corners, a tesseract has $2^4 = 16$ corners, a 5D hypercube has $2^5 = 32$ corners, etc.

	Point	Line	Square	Cube	Tesseract	5D Cube	N-D Cube
Corners	1	2	4	8	16	32	2^N

In the notation x^n, which reads x raised to the power of n, n is called an *exponent*. Mathematically, x^n means to multiply n x's together. For example, $3^4 = 3 \cdot 3 \cdot 3 \cdot 3 = 81$. If $n = 2$, x is said to be *squared*; and if $n = 3$, x is said to be *cubed*. It is easy to see that $x^m x^n = x^{m+n}$. This works for negative exponents, provided that $x^{-n} = 1/x^n$. Since $x^m x^{-m} = x^0$ must be 1, any number raised to the zero power is unity: $x^0 = 1$.

An N-dimensional cube has N sets of 2^{N-1} parallel edges – i.e. $N(2^{N-1})$ edges all together. A point has no edges in accordance with $0(2^{-1}) = 0$, a line segment is its own edge as $1(2^0) = 1$, a square has $2(2^1) = 4$ edges, a cube has $3(2^2) = 12$ edges, a tesseract has $4(2^3) = 32$ edges, a 5D hypercube has $5(2^4) = 80$ edges, and so on. Another way to put it is: A square has 2 pairs of parallel edges, a cube has 3 sets of 4 parallel edges, a tesseract has 4 sets of 8 parallel edges, a 5D hypercube has 5 sets of 16 parallel edges, etc.

	Line	Square	Cube	Tesseract	5D Cube	6D Cube	N-D Cube
Edges	1	4	12	32	80	192	$N \cdot 2^{N-1}$

An N-dimensional cube has $N(N-1)/2!$ sets of 2^{N-2} parallel, square faces – $N(N-1)2^{N-2}/2!$ squares in all. A cube has $(3 \cdot 2/2)2^1 = 6$ faces, a tesseract has $(4 \cdot 3/2)2^2 = 24$ squares, a 5D hypercube has $(5 \cdot 4/2)2^3 = 80$ squares, etc. A cube has 3 sets of 2 parallel faces, a tesseract has 6 sets of 4 parallel squares, a 5D hypercube has 10 sets of 8 parallel squares, a 5D hypercube has 5 sets of 16 parallel squares, etc.

	Square	Cube	Tesseract	5D Cube	6D Cube	N-D Cube
Squares	1	6	24	80	240	$\dfrac{N(N-1)2^{N-2}}{2!}$

For a nonnegative integer n, $n!$ – which reads as, "n *factorial*" – means n times $(n-1)$ times $(n-2)$ and so on, ending with multiplication by 1. For example, $4! = 4 \cdot 3 \cdot 2 \cdot 1 = 24$. Special cases include $1! = 1$ and $0! = 1$.

An N-dimensional cube contains $N(N-1)(N-2)/3!$ sets of 2^{N-3} parallel cubes – a total of $N(N-1)(N-2)2^{N-3}/3!$ cubes. A tesseract is bounded by $(4 \cdot 3 \cdot 2/6)2^1 = 8$ cubes, a 5D hypercube contains $(5 \cdot 4 \cdot 3/6)2^2 = 40$ cubes, and so forth.

	Cube	Tesseract	5D Cube	6D Cube	N-D Cube
Cubes	1	8	40	160	$\dfrac{N(N-1)(N-2)2^{N-3}}{3!}$

An N-dimensional cube contains a total of $N(N-1)(N-2)(N-3)2^{N-4}/4!$ tesseracts. A 5D hypercube is bounded by $(5 \cdot 4 \cdot 3 \cdot 2/24)2^1 = 10$ tesseracts, for example.

	Tesseract	5D Cube	6D Cube	N-D Cube
Tesseracts	1	10	60	$\dfrac{N(N-1)(N-2)(N-3)2^{N-4}}{4!}$

Several patterns can be seen by compiling all of this data into a single table:

	Point	Line	Square	Cube	Tesseract	5D Cube	6D Cube
Corners	1	2	4	8	16	32	64
Edges		1	4	12	32	80	192
Squares			1	6	24	80	240
Cubes				1	8	40	160
Tesseracts					1	10	60
5D Cubes						1	12

There are diagonal patterns as well as horizontal patterns. The leftmost diagonal consists of 1's, while the second runs $2, 4, 6, 8 \ldots$ The patterns for the third and fourth diagonal sequences, $4, 12, 24, 40 \ldots$ and $8, 32, 80, 160 \ldots$, respectively, are not as obvious. Nevertheless, we can develop formulas for the various patterns.

The sub-cube patterns can be explained concisely using combinatorial notation. The combinatorial, written $\begin{pmatrix} N \\ M \end{pmatrix}$, reads " N

81

choose M," meaning, out of N objects, how many distinct ways are there to select just M of them, where $M \leq N$. It turns out that the answer is given in terms of factorials:

$$\binom{N}{M} = \frac{N!}{M!(N-M)!}$$

For example, given 4 independent coordinates (x, y, z, and w), there are $\binom{4}{3} = \frac{4!}{3!\,1!} = 4$ ways to choose three of them. These correspond to the 4 orthogonal 3D hyperplanes: xyz, wxy, zwx, and yzw. In 4D, there are $\binom{4}{2} = \frac{4!}{2!\,2!} = 6$ ways to choose two independent coordinates, which correspond to the 6 orthogonal planes: xy, zx, xw, yz, zw, and wy. The are $\binom{4}{1} = \frac{4!}{1!\,3!} = 4$ ways to choose one out of the four independent coordinates; these are the x-, y-, z-, and w-axes. Finally, there is just $\binom{4}{4} = \frac{4!}{4!\,0!} = 1$ way to choose all four independent coordinates.

In general, the formula to compute the number of M-dimensional sub-cubes contained in an N-dimensional cube is $2^{N-M}\binom{N}{M}$. For example, the number of faces ($M = 2$) contained in a 5D hypercube ($N = 5$) equals $(2^3)\frac{5!}{2!\,3!} = 80$.

Going across on the sub-cube pattern chart corresponds to an increase in N, the number of dimensions of the N-dimensional cube. Reading down corresponds to increasing M, the dimensionality of the M-dimensional sub-cube. Selecting any numerical entry in the table, corresponding to an N-dimensional cube and an M-dimensional sub-cube, the value of the first cell to its right is given by multiplying the first cell's value by $2(N+1)/(N-M+1)$. For example, the entry 12 corresponds to $N = 3$ and $M = 1$. The multiplicative factor is $2 \cdot 4/3 = 8/3$, and $12 \cdot 8/3 = 32$, which agrees with the next cell over. This entry 32 corresponds to $N = 4$ and $M = 1$, for which the multiplicative factor is $2 \cdot 5/4 = 5/2$; multiplying this by 32 gives 80, which is the next cell over. Observe that, while going across, N increases by one unit per cell while M remains constant.

The multiplicative factor for going one cell downward is $\dfrac{N-M}{2(M+1)}$. Going one cell downward, M increases by one unit while N remains constant. Starting with the entry 8, for which $N=3$ and $M=0$, the first multiplicative factor is $3/2$, which transforms the 8 into 12; the second multiplicative factor, $1/2$, turns this into a 6; and the last multiplicative factor for this column, $1/6$, changes this into a 1.

The diagonal sub-cube pattern is thus seen to follow a combination of the row and column formulas, with one subtlety – the value of N increases by one going one cell to the right, before increasing M by one unit going down one cell. Going diagonally down to the right one cell, the multiplicative factor is $(N+1)/(M+1)$. The first diagonal corresponds to $N=M$, for which the multiplicative factor is simply 1; this is why the first diagonal consists of all 1's. For the second diagonal, $N=M+1$, so the first factor, starting with the $N=1$, $M=0$ cell, is 2, which makes the 2 become a 4; the next factor is $3/2$, transforming the 4 into a 6. The third diagonal, for which $N=M+2$, gives factors of 3, 2, $5/3$, etc.

For an N-dimensional cube: N edges meet at each corner, $N-1$ squares share every edge, $N-2$ cubes are joined to each square face, $N-3$ tesseracts meet at every cube, etc; $\binom{N}{2}$ squares meet at each corner, $\binom{N-1}{2}$ cubes share each edge, $\binom{N-2}{2}$ tesseracts meet at every square, etc.; $\binom{N}{3}$ cubes meet at each corner, $\binom{N-1}{3}$ share every edge, $\binom{N-2}{3}$ 5D hypercubes are joined at each square, etc.; and so on. For example, for the 5D hypercube, for which $N=5$, 5 edges intersect at every corner, 4 squares share each edge, each square is shared by 3 different cubes, and each cube is one of the bounding cubes for 2 different tesseracts. In general, the number of M-dimensional sub-cubes meeting at every L-dimensional sub-cube in an N-

dimensional cube is given by $\begin{pmatrix} N-L \\ M-L \end{pmatrix}$, where $L \leq M \leq N$. For

example, $\dfrac{4!}{3!\,1!} = 4$ tesseracts $(M=4)$ meet at each edge $(L=1)$ of a

5D hypercube $(N=5)$.

	Square	Cube	Tesseract	5D Cube	6D Cube
Edges per Corner	2	3	4	5	6
Squares per Corner		3	6	10	15
Squares per Edge		2	3	4	5
Cubes per Corner			4	10	20
Cubes per Edge			3	6	10
Cubes per Square			2	3	4
Tesseracts per Corner				5	15
Tesseracts per Edge				4	10
Tesseracts per Square				3	6
Tesseracts per Cube				2	3

Puzzle 3.7: How many cubes meet at each corner of a 7D hypercube?

The Schläfli symbol for an N-dimensional cube progresses as follows: $\{4\}$ for a square since it is a four-sided polygon; $\{4,3\}$ for a cube because its faces have four sides and three faces meet at every vertex; $\{4,3,3\}$ for a tesseract, where the first two values are unchanged and the last 3 indicates that three cubes meet at each edge; $\{4,3,3,3\}$ for a 5D hypercube, where the last 3 now corresponds to the number of tesseracts meeting at each square; etc. In general, it has the form $\{4,3,3,...,3\}$ for an N-dimensional cube, where $N-2$ three's follow the leading four.

The $(N-1)$-dimensional cross section of an N-dimensional cube can be an $(N-1)$-dimensional polytope with as many as $2N$ sides. For example, the 2D cross section of a cube – i.e. the intersection of a plane and a cube – can be a polygon with as many as 6 sides (which is a hexagon). Other possible 2D polygonal cross sections for a cube include a triangle, quadrilateral, and pentagon; and it is also possible for the 2D cross section of a cube to be a point or a line segment. The 3D cross section of a tesseract can be a polyhedron with as many as 8 sides (an octahedron).

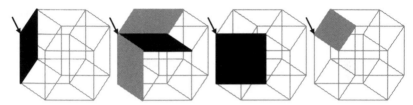

6 squares meeting at 1 corner of a tesseract

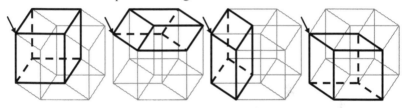

4 cubes meeting at 1 corner of a tesseract

3 squares meeting at 1 edge of a tesseract

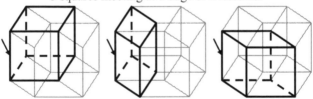

3 cubes meeting at 1 edge of a tesseract

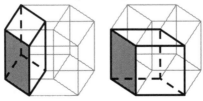

2 cubes meeting at 1 square of a tesseract

3.3 Anatomy of a Pentachoron

A tetrahedron is a polyhedron consisting of four sides, each of which is a triangle. For regular tetrahedra, each triangle is equilateral. The Schläfli symbol for the tetrahedron is {3,3}, since three three-sided faces meet at each vertex. (Notice that tetrahedra differ from the ancient pyramids in that the base is triangular rather than square.)

tetrahedra

The pentachoron (aka pentatope, 5-cell, 4D simplex, pentahedroid, and hypertetrahedron) is the 4D generalization of the tetrahedron. Regular pentachora are bound by five regular tetrahedra. The Schläfli symbol for the pentachoron is {3,3,3}, where the final 3 indicates that three tetrahedra meet at each edge.

The N-dimensional simplex is the simplest N-dimensional polytope in the sense that it is the N-dimensional polytope bounded by the fewest $(N-1)$-dimensional polytopes. The pentachoron is the simplex for which $N=4$.

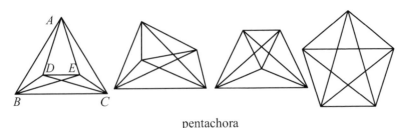

pentachora

With 3 vertices, the triangle is the simplest polygon. Similarly, with 4 vertices, the tetrahedron is the simplest polyhedron; it is a 3D simplex. The pentachoron is a 4D simplex, containing 5 vertices. The names of these polytopes derive, in part, from the number of vertices: The prefix *tri-* means three, *tetra-* means four, and *penta-* means five.

86

Each pair of points in a simplex is joined by an edge. The triangle has 3 edges and the tetrahedron has 6 edges. The pentachoron has 10 edges:

$$\overline{AB} \quad \overline{AC} \quad \overline{AD} \quad \overline{AE} \quad \overline{BC}$$

$$\overline{BD} \quad \overline{BE} \quad \overline{CD} \quad \overline{CE} \quad \overline{DE}$$

Every group of three points in a simplex forms a triangle. Thus, the tetrahedron is bounded by 4 triangular faces. The pentachoron contains 10 triangles (called *ridges*):

$$\triangle ABCA \quad \triangle ABDA \quad \triangle ABEA \quad \triangle ACDA \quad \triangle ACEA$$

$$\triangle ADEA \quad \triangle BCDB \quad \triangle BCEB \quad \triangle BDEB \quad \triangle CDEC$$

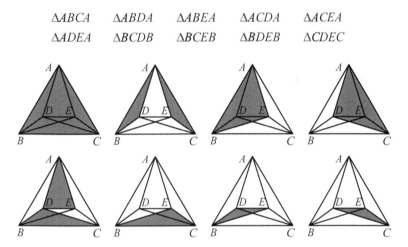

10 triangles of a pentachoron

A pentachoron is bounded by 5 tetrahedra (called *facets*, not to be confused with the *faces* of polyhedra):

$$ABCD \quad ABCE \quad ABDE \quad ACDE \quad BCDE$$

For a triangle, 2 lines meet at each vertex. For a tetrahedron, 3 edges intersect at each vertex, 3 triangles join together at each vertex, and 2 triangles meet at every edge. For a pentachoron, 4 lines merge at each vertex, 6 triangles meet at every vertex, 3 triangles join at each edge, 4 tetrahedra meet at each vertex, 3 tetrahedra join at every edge, and 2 tetrahedra share each triangle.

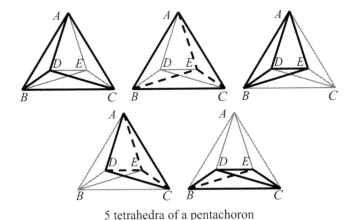

5 tetrahedra of a pentachoron

Puzzle 3.8: What are the possible polygonal cross sections of a tetrahedron? What are the possible polyhedral cross sections of a pentachoron?

3.4 Simplex Patterns

The simplex is an N-dimensional generalization of the tetrahedron, and is the simplest polytope for a given dimensionality N in the sense that it has the fewest number of vertices possible. N-dimensional simplexes (alternatively, simplices) are bound by $(N-1)$-dimensional simplexes. The 0D simplex is a point, the 1D simplex is a line segment, the 2D simplex is a triangle, the 3D simplex is a tetrahedron, the 4D simplex is a pentachoron, and in higher dimensions it is convenient to just specify the dimensionality – e.g. the 8D simplex. Notice that the pentachoron is a 4D simplex, and that the 5D simplex has six corners – i.e. the *penta*- in pentachoron refers to the number of vertices, not the number of dimensions.

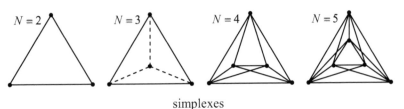

simplexes

An N-dimensional simplex has $N+1$ vertices. The triangle has 3 vertices, the tetrahedron has 4 vertices, the pentachoron has 5 vertices, the 5D simplex has 6 vertices, etc.

	0D	1D	2D	3D	4D	5D	6D	N-D Simplex
Vertices	1	2	3	4	5	6	7	$N+1$

An N-dimensional simplex has $(N+1)N/2!$ edges. The triangle has $3\cdot2/2=3$ edges, the tetrahedron has $4\cdot3/2=6$ edges, the pentachoron has $5\cdot4/2=10$ edges, the 5D simplex has $6\cdot5/2=15$ edges, and so on.

	1D	2D	3D	4D	5D	6D	7D	N-D Simplex
Edges	1	3	6	10	15	21	28	$(N+1)N/2!$

An N-dimensional simplex contains $(N+1)N(N-1)/3!$ triangles. The tetrahedron is bounded by $4\cdot3\cdot2/6=4$ triangles, the pentachoron has $5\cdot4\cdot3/6=10$ triangles, the 5D simplex has $6\cdot5\cdot4/6=20$ triangles, etc.

	2D	3D	4D	5D	6D	7D	8D	N-D Simplex
Triangles	1	4	10	20	35	56	84	$(N+1)N(N-1)/3!$

An N-dimensional simplex consists of $(N+1)N(N-1)(N-2)/4!$ tetrahedra. The pentachoron is bounded by $5\cdot4\cdot3\cdot2/24=5$ tetrahedra, the 5D simplex contains $6\cdot5\cdot4\cdot3/24=15$ tetrahedra, and so on.

	3D	4D	5D	6D	7D	8 D	N-D Simplex
Tetrahedra	1	5	15	35	70	126	$(N+1)N(N-1)(N-2)/4!$

Several simplex patterns are seen when this data is tabulated together:

	0D	1D	2D	3D	4D	5D	6D	7D	8D
Vertices	1	2	3	4	5	6	7	8	9
Edges		1	3	6	10	15	21	28	35
Triangles			1	4	10	20	35	56	84
Tetrahedra				1	5	15	35	70	126
Pentachora					1	6	21	56	126
5D Simplexes						1	7	28	84
6D Simplexes							1	8	36
7D Simplexes								1	9
8D Simplexes									1

Puzzle 3.9: What are the diagonal patterns for this table? Use this to predict the number of pentachora in a 9D simplex.

The number of M-dimensional sub-simplexes in an N-dimensional simplex (so $M \leq N$) is given by $\binom{N+1}{M+1}$. Statistically, this says, out of $N+1$ corners, how many ways there are to use $M+1$ of the corners. For example, the number of triangles ($M = 2$) in a pentachoron ($N = 4$) corresponds to the number of ways to use 3 of the 5 corners, which is $\binom{5}{3} = \dfrac{5!}{3!\,2!} = 10$. (The reason for the $+1$'s is that there are $N+1$ corners in an N-dimensional simplex.) The entries in the simplex table also appear as coefficients in the binomial expansion – i.e. the expansion of $(x+y)^N$:

$$(x+y)^0 = 1$$

$$(x+y)^1 = x+y$$

$$(x+y)^2 = x^2 + 2xy + y^2$$

$$(x+y)^3 = x^3 + 3x^2 y + 3xy^2 + y^3$$

$$(x+y)^4 = x^4 + 4x^3 y + 6x^2 y^2 + 4xy^3 + y^4$$

$$(x+y)^5 = x^5 + 5x^4 y + 10x^3 y^2 + 10x^2 y^3 + 5xy^4 + y^5$$

$$(x+y)^6 = x^6 + 6x^5 y + 15x^4 y^2 + 20x^3 y^3 + 15x^2 y^4 + 6xy^5 + y^6$$

$$(x+y)^7 = x^7 + 7x^6 y + 21x^5 y^2 + 35x^4 y^3 + 35x^3 y^4 + 21x^2 y^5 + 7xy^6 + y^7$$

Puzzle 3.10: What is the coefficient of $x^5 y^3$ in $(x+y)^8$?

For a triangle, 2 line segments meet at each vertex. For a tetrahedron, 3 edges intersect at each vertex and 3 triangles meet at each vertex, while 2 triangles share every edge. For a polychoron, 4 edges intersect at each vertex, 6 triangles meet at each vertex, 3 triangles share every edge, 4 tetrahedra meet at each vertex, 3 tetrahedra intersect at every edge, and 2 tetrahedra share every triangle. The pattern of intersections is the same as that of the hypercube.

	2D	3D	4D	5D	6D	7D	8D
Edges per Vertex	2	3	4	5	6	7	8
Triangles per Vertex		3	6	10	15	21	28
Triangles per edge		2	3	4	5	6	7
Tetrahedra per Vertex			4	10	20	35	56
Tetrahedra per Edge			3	6	10	15	21
Tetrahedra per Triangle			2	3	4	5	6
Polychora per Vertex				5	15	35	70
Polychora per Edge				4	10	20	35
Polychora per Triangle				3	6	10	15
Polychora per Tetrahedron				2	3	4	5

The Schläfli symbol for an N-dimensional simplex progresses as follows: {3} for a triangle since it is a three-sided polygon; {3,3} for a tetrahedron, where the second three indicates that three triangles meet at every vertex; {3,3,3} for a pentachoron, where the last 3 indicates that three tetrahedra meet at each edge; etc. The

91

Schläfli symbol for an N-dimensional simplex has the form $\{3,3,3,...,3\}$, where there are $N-1$ three's all together.

3.5 Other Polytopes

A dodecahedron is a polyhedron bounded by 12 pentagons. For a regular dodecahedra, the pentagons are also regular. The Schläfli symbol for the dodecahedron is $\{5,3\}$, since three five-sided faces meet at each vertex. A dodecahedron has 20 vertices, 30 edges, and 12 faces. 3 edges meet at each vertex, 3 faces meet at each vertex, and 2 faces meet at each edge.

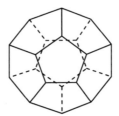

a regular dodecahedron

One way to draw a dodecahedron is to first draw large circle. Mark 10 points on the circle 36° apart (one-tenth of 360°, which corresponds to a complete circle), as measured from the center. Connect these 10 points to form a regular decagon, which will be the 2D outline of the dodecahedron. Draw a second circle, concentric and inside the first circle to help draw the front and rear pentagon faces. Mark the corresponding 10 points on this inner circle. Connect every other point to form one pentagon, and connect the remaining points to form the other pentagon; used dashed lines for the rear pentagon. Connect each point of the front pentagon to the corresponding point of the large decagon, and similarly for the rear pentagon, using dashed lines for the rear.

A regular octahedron is bounded by 8 equilateral triangles in a diamond shape. An octahedron has 6 vertices, 12 edges, and 8 faces. 4 edges and 4 faces meet at each vertex, and 2 faces meet at each edge. The Schläfli symbol for the octahedron is $\{3,4\}$.

A regular icosahedron is bound by 20 equilateral triangles. An icosahedron has 12 vertices, 30 edges, and 20 faces. 5 edges and 5 faces meet at each vertex, and 2 faces meet at each edge. The Schläfli symbol for the icosahedron is {3,5}. One view of an icosahedron can be drawn from the outline of a decagon.

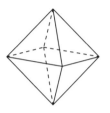 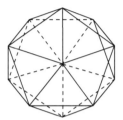

a regular octahedron   a regular icosahedron

The five regular convex polyhedra (aka the five Platonic solids) include the tetrahedron, cube, octahedron, dodecahedron, and icosahedron. There are also four regular concave (stellated) polyhedra. In addition to having regular faces, the vertex figures (achieved by cutting off a corner at the midpoints of the edges) of a polyhedron are also regular.

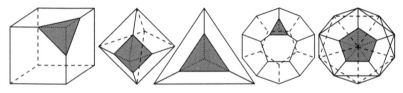

vertex figures

The four regular concave polyhedra are created by starting with a regular convex polyhedron and appending a spire to each face by extending the planes of the old adjacent faces. These concave polyhedra are thus stellated, appearing as 3D star-shapes.

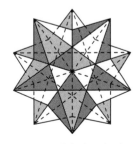

small stellated dodecahedron

For every regular convex polytope, there is a dual regular convex polytope related by the vertex figures. The cube and octahedron are dual to each other. Their Schläfli symbols are {4,3} and {3,4}, respectively. The vertex figure of the cube is an equilateral triangle, while the vertex figure of the octahedron is a square. A cube has 8 vertices, 12 edges, and 6 faces, while an octahedron has 6 vertices, 12 edges, and 8 faces. The dodecahedron and icosahedron are similarly dual to one another. Their Schläfli symbols are {5,3} and {3,5}, respectively. Their vertex figures are similarly swapped: equilateral triangles for the dodecahedron and pentagons for the icosahedron. The dodecahedron has 20 vertices, 30 edges, and 12 faces, and the icosahedron has 12 vertices, 30 edges, and 20 faces. The tetrahedron, with Schläfli symbol {3,3} and triangular vertex figures, is said to be self-dual.

There are six regular convex polychora and ten regular concave (stellated) polychora. The six regular convex polychora include the pentachoron, the tesseract, the hexadecachoron, the icositetrachoron, the hecatonicosahedron, and the hexacosichoron.

A regular hexadecachoron (aka 16-cell or hyperoctahedron) has 8 vertices, 24 edges, 32 triangular faces, and 16 bounding tetrahedra. Its Schläfli symbol is {3,3,4}. It is one way to generalize the octahedron: The hexadecachoron can be drawn by first drawing a square in the xy plane that is rotated 45° about the z-axis, and then drawing two octahedra sharing this square base, one that extends along $\pm z$ and the other along $\pm w$. The hexadecachoron is dual to the tesseract, which has Schläfli symbol {4,3,3}, 16 vertices, 32 edges, 24 square faces, and 8 bounding cubes. While a tesseract has 3 cubes per edge, a hexadecachoron has 4 tetrahedra per edge. The tesseract is a 4D hypercube, and the hexadecachoron is a 4D cross polytope; the hypercube and cross polytope are two different ways to generalize a square to higher dimensions.

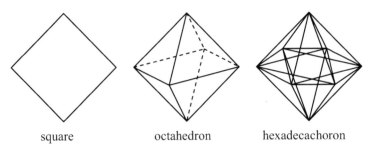

square   octahedron   hexadecachoron

The unit cross polytope has vertices at $(\pm 1,0,0,...,0)$, $(0,\pm 1,0,...,0)$, $(0,0,\pm 1,...,0)$, ..., $(0,0,0,...,\pm 1)$; all vertices are connected with edges except for opposite vertices. The N-dimensional cross polytope is a point in 0D, a line segment in 1D, a square in 2D, an octahedron in 3D, a hexadecachoron in 4D, etc.

Puzzle 3.11: How many squares are present in a hexadecachoron, and what is their significance?

Puzzle 3.12: Find the number of vertices, edges, triangular faces, tetrahedra, and bounding hexadecachora in a 5D cross polytope.

A regular icositetrachoron (aka 24-cell or hyperdiamond) has 24 vertices, 96 edges, 96 triangular faces, and 24 bounding octahedra. With 3 octahedra meeting at each edge, its Schläfli symbol is $\{3,4,3\}$. With the 4 in the middle, it is neither dual to the tesseract nor the hexadecachoron, even though their Schläfli symbols have the same digits in different order. In fact, the icositetrachoron is self-dual: Notice that its Schläfli symbol is its own reverse, as are the number of vertices, edges, faces, and octahedra. It does not have a direct analogy in 3D, though it can be constructed by adding 8 hyperpyramids to the 8 cubes bounding a tesseract. Whereas a 3D pyramid has a square base, a 4D hyperpyramid has a cubic base –i.e. the apex connects to all 8 corners of the cube. This construction is similar to the 3D rhombic dodecahedron, which is made by adding 6 pyramids to the 6 square faces bounding a cube.

The hecatonicosachoron (aka 120-cell or hyperdodecahedron) has 600 vertices, 1200 edges, 720 pentagon faces, and 120 bounding dodecahedra. Its Schläfli symbol is $\{5,3,3\}$. The hecatonicosachoron is the 4D generalization of the dodecahedron.

The hexacosichoron (aka 600-cell or hypericosahedron) is dual to the hecatonicosachoron, with 120 vertices, 720 edges, 1200 triangular faces, 600 bounding tetrahedra, and Schläfli symbol $\{3,3,5\}$. The hexacosichoron is the 4D generalization of the icosahedron, which is dual to the dodecahedron.

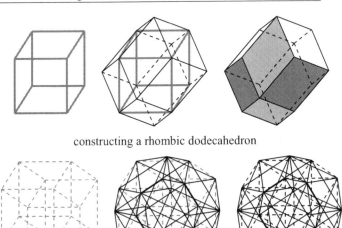

constructing a rhombic dodecahedron

constructing an icositetrachoron

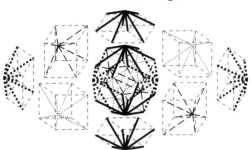
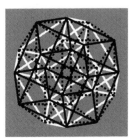

8 hyperpyramids for 8 cubes bounding a tesseract

an icositetrachoron divided into 3 tesseracts

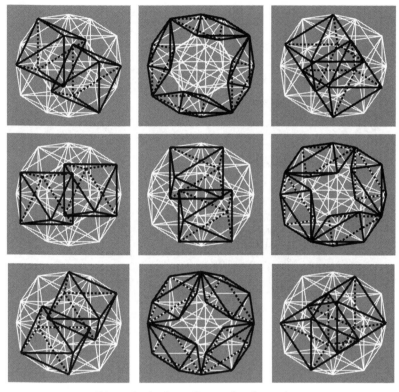

24 octahedra bounding an icositetrachoron

Further Reading

See Coxeter's work [A6] for a classic introduction to polytopes. Rudy Rucker discusses hypercubes and pentachora in his classic introduction to the fourth dimension [A3]. Many websites also provide information on 4D objects, such as www.mathworld.com.

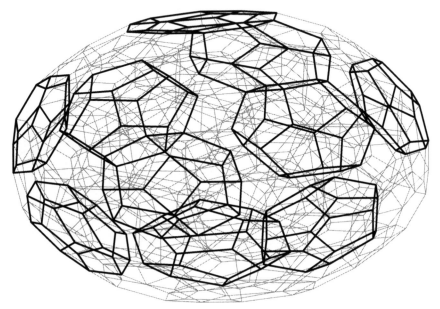

a hecatonicosachoron (with 10 of the 120 bounding dodecahedra highlighted)

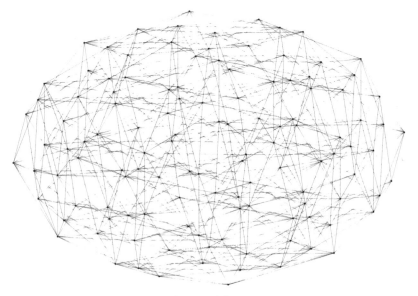

a hexacosichoron

4 Curved Hypersurfaces

The curved hypersurfaces studied here will be useful for understanding compactification in Chapter 8 and help to visualize higher-dimensional structures – such as what a 4D planet would look like. We will begin with an introductory review of common curves and surfaces, like parabolas and ellipsoids, and then we will generalize these to hypersurfaces. A method has been developed for displaying 4D graphs that is similar to the contours used to depict 3D graphs, so studying the surfaces of the 2D plots will be useful. Some equations are provided for those interested in understanding mathematically the origins of the hypersurfaces or with the desire to write computer programs to graph higher-dimensional objects. There is also a little math in the section on hypersphere patterns. The math is easily skipped by those who prefer to focus on the concepts. At a minimum, an effort should be made to understand the basic 4D structures – e.g. the distinction between a spherinder and a cubinder.

4.0 Curves

An object traveling along a curved track experience one degree of freedom: The only choice for the direction of motion is forward or backward. In this sense, a curve is 1D. However, the curve itself does wind its way through two or more dimensions. The design and construction of a roller coaster is 3D, yet the riders experience 1D motion.

A line is a 1D path that does not curve, and so does not wind its way through a higher-dimensional space. In 2D, the equation for a straight line is $y = mx + b$, where m is the slope of the line and b is the y-intercept. The slope can be computed from the (x, y) coordinates of any two points on the line by dividing the difference in y-values by the difference in x-values: $m = (y_2 - y_1)/(x_2 - x_1)$. The slope is thus the rise $(y_2 - y_1)$ over the run $(x_2 - x_1)$. The effect of changing the slope is to change the steepness of the line. A horizontal line has no steepness – i.e. $m = 0$. A line that slants upwards has positive slope, while a line that slants downward has negative slope. The y-intercept is the value of y for which the line intersects the y-axis, which is the value of y for which $x = 0$.

Interestingly, in 3D, the equation $y = mx + b$ actually represents a plane, not a line. The reason is that the equation $y = mx + b$ does not include a z ; thus, y equals $mx + b$ for every value of z. This infinite number of lines (each line corresponding to a different value of z) amounts to a plane. In 3D, the equation for a line can instead by specified as the intersection of two planes by giving two equations, one for each plane. In 4D, the equation $y = mx + b$ is neither a line nor a plane – it is a 3D subspace (it is a 3D space, which is a subspace of the 4D space). The intersection of two 3D subspaces is a plane, so a plane can be specified in 4D by giving two equations, one for each 3D subspace; three equations together are needed to specify a line.

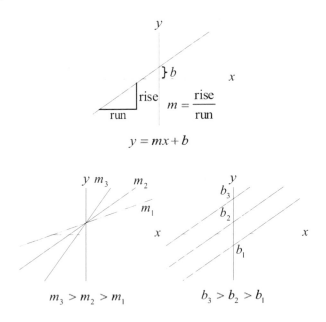

Consider two coaxial cones joined vertex-to-vertex, where one cone is up-side-down compared to the other. The intersection of these coins with a plane yields the following cross sections: A point, a line, a circle, an ellipse, a parabola, or a hyperbola. It turns out that these latter five cross sections, termed the *conic sections*, are the possible orbits of a planet, moon, or other satellite (in the two-body approximation). The cross section is a point if the plane intersects only the vertex, and it is a line if the plane is tilted so as to match the slope of the cone and also passes through the vertex; it can be a pair of lines if the axis of the cones lies in the plane.

A circular cross section is obtained when a plane perpendicular to the axis intersects one of the cones. A circle is a 2D locus of points equidistant from a common point, which defines the center of the circle. In 2D, the equation for a circle centered about the origin is $x^2 + y^2 = R^2$, where R is the radius of the circle. The same equation in 3D defines a right-circular cylinder since $x^2 + y^2$ equals R^2 for every value of z. A circle can be specified in 3D by giving a pair of equations.

If the plane that produces a circular cross section is tilted a little, the cross section of the cone becomes an ellipse. The cross section will remain an ellipse until the tilt of the plane reaches the slope of the cone. An ellipse is the locus of points for which the sum of the distances to two different points is a constant; each point is called a *focus*. In 2D, the equation for an ellipse centered about the origin is $\left(\dfrac{x}{a}\right)^2 + \left(\dfrac{y}{b}\right)^2 = 1$, where the larger of a and b is termed the *semimajor axis* and the smaller is called the *semiminor axis*.

If the plane matches the slope of the cone, but does not pass through the vertex, the cross section is a parabola. A parabola is the locus of points for which the distance to a fixed point equals the distance to a fixed line. The fixed point is called the *focus* and the fixed line is termed the *directrix*. In 2D, the equation for a parabola symmetric about the y-axis is $y = ax^2 + b$, where $a/4$ is the distance from the focus to the vertex (the point of the parabola lying on the symmetry axis, which lies halfway between the focus and the directrix) and b is the y-intercept.

If the plane is parallel to the axis of the cones but does not include the axis, the cross section is a pair of hyperbolas. A hyperbola is the locus of points for which the distance to one fixed point minus the distance to a second fixed point is constant. Again, these fixed points are called *foci*. In 2D, the equation for a hyperbola with its foci lying on the x-axis is $\left(\dfrac{y}{b}\right)^2 - \left(\dfrac{x}{a}\right)^2 = 1$. The hyperbola has two asymptotes, which are the lines $y = \pm bx/a$. The hyperbola approaches these asymptotes closer and closer as it extends, but never quite reaches them.

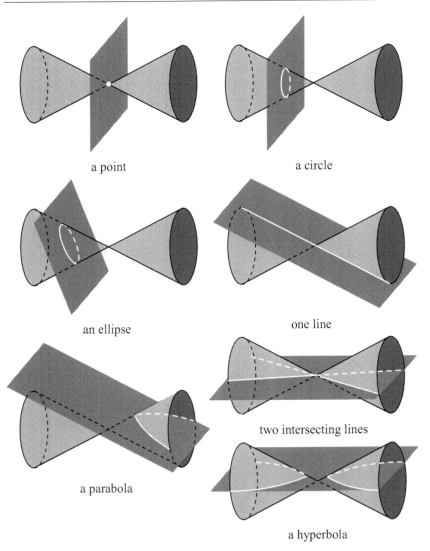

a point

a circle

an ellipse

one line

a parabola

two intersecting lines

a hyperbola

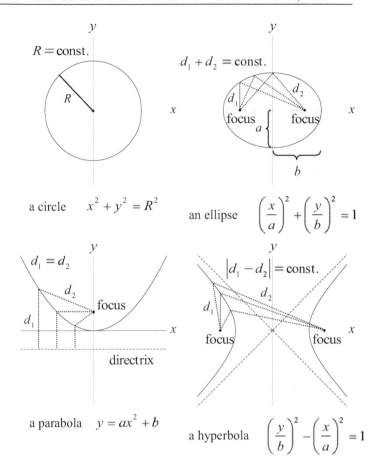

a circle $\quad x^2 + y^2 = R^2$

an ellipse $\quad \left(\dfrac{x}{a}\right)^2 + \left(\dfrac{y}{b}\right)^2 = 1$

a parabola $\quad y = ax^2 + b$

a hyperbola $\quad \left(\dfrac{y}{b}\right)^2 - \left(\dfrac{x}{a}\right)^2 = 1$

4.1 Surfaces

Although a surface may curve through a 3D (or higher) space, a surface is 2D in the sense that an object confined to motion along the surface has only two degrees of freedom.

A plane is a flat surface. In 3D, the equation for a plane is $z = ax + by + c$, which is similar in form to the equation for a line in 2D. In fact, the equation for a line in 2D represents a plane in 3D. The constants a and b affect the slope of the plane, and c is the z-intercept – i.e. the value of z for which the plane intersects the z-axis.

103

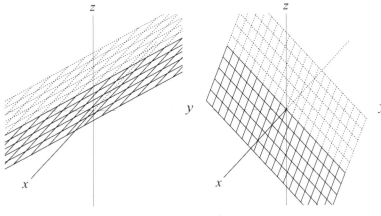

planes $z = ax + by + c$

A sphere is a 3D locus of points equidistant from a common point – the center. In 3D, the equation for a sphere centered about the origin is $x^2 + y^2 + z^2 = R^2$, where R is the radius of the sphere. In the diagrams that follow, the spheres and other surfaces appear distorted because they are not drawn in perspective – as you would view the surface of a ball with your eye.

An ellipsoid is one way to generalize the ellipse to 3D. In 3D, the equation for an ellipsoid centered with its symmetry axes along the coordinate axes is $\left(\dfrac{x}{a}\right)^2 + \left(\dfrac{y}{b}\right)^2 + \left(\dfrac{z}{c}\right)^2 = 1$. Unless two or more of the constants are equal, the cross sections of the ellipsoid in the xy, yz, and zx planes are ellipses.

The spheroid is a special case of the ellipsoid, in which two of the three constants are equal. A spheroid looks like a sphere that has been flattened at opposite poles. If $b = c$, for example, then the cross section in the yz plane is a circle, while the cross sections in the xy and zx planes are ellipses. If $b = c$, $|a| > |b|$ makes a prolate spheroid, while $|a| < |b|$ creates an oblate spheroid. The distinction determines whether or not the circular cross section in the yz plane is greater or smaller than the elliptical cross sections in the xy and zx planes.

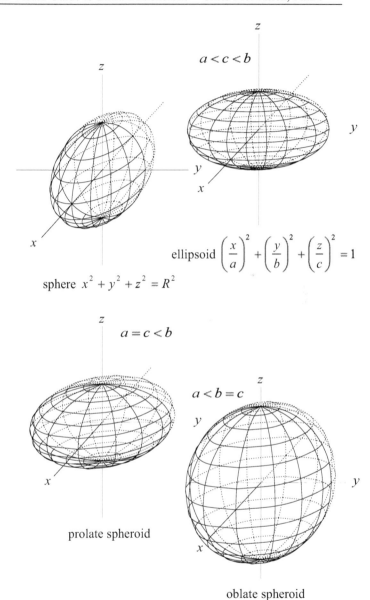

$a < c < b$

ellipsoid $\left(\dfrac{x}{a}\right)^2 + \left(\dfrac{y}{b}\right)^2 + \left(\dfrac{z}{c}\right)^2 = 1$

sphere $x^2 + y^2 + z^2 = R^2$

$a = c < b$

$a < b = c$

prolate spheroid

oblate spheroid

A right-circular cylinder is the 3D locus of points equidistant from a line, which defines the axis of the cylinder. The equation for a right-circular cylinder concentric with the z-axis is $x^2 + y^2 = R^2$.

This same equation is a circle in 2D. In 3D, $x^2 + y^2$ equals R^2 for every value of z, creating a cylinder. The cross section of a cylinder need not be a circle, in general. For example, a right-elliptical cylinder has elliptical cross section. Other possibilities include parabolic and hyperbolic cylinders. A cylinder need not even be right – instead, it may be slanted, but it will still have constant cross section (unlike a cone).

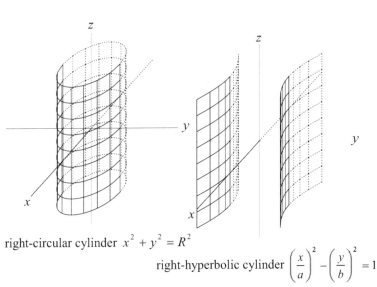

right-circular cylinder $x^2 + y^2 = R^2$

right-hyperbolic cylinder $\left(\dfrac{x}{a}\right)^2 - \left(\dfrac{y}{b}\right)^2 = 1$

A right-circular cone is a like a right-circular cylinder, except that the radius of the circular cross section varies linearly, tapering to a point at one end; whereas the cross section of a right-circular cylinder has constant radius. The equation for a right-circular cone with its vertex at the origin and axis along the z-axis is $x^2 + y^2 = az^2$, where the constant a affects the angle of the cone. This equation actually yields a double-cone – one for positive and one for negative values of z. In general, a cone may also have a slant or a cross section other than a circle.

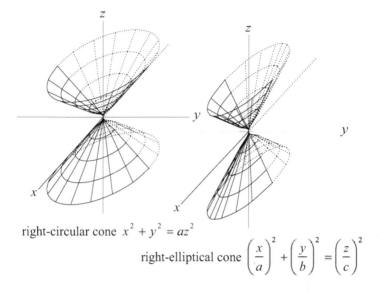

right-circular cone $x^2 + y^2 = az^2$

right-elliptical cone $\left(\dfrac{x}{a}\right)^2 + \left(\dfrac{y}{b}\right)^2 = \left(\dfrac{z}{c}\right)^2$

A paraboloid is a generalization of the parabola to 3D. The circular paraboloid resembles a curved right-circular cone, with notable differences being that it looks parabolic rather than triangular from the side view and does not stand on its mirror image. The circular paraboloid is shaped like a bowl. The equation for a circular paraboloid with its vertex at the origin and symmetry axis along the positive z-axis is $z = ax^2 + ay^2$. Its cross section parallel to the xy plane is a variable-sized circle, while its cross sections in the yz and zx planes are parabolas; its parabolic cross sections in the yz and zx planes distinguish it from the right-circular cone. An elliptic paraboloid is created by making the coefficients of x^2 and y^2 differ – i.e. $z = ax^2 + by^2$, where a and b have the same sign. The cross section parallel to the xy plane becomes a variable-sized ellipse. The equation $z = ax^2 - by^2$ results in a hyperbolic paraboloid if a and b have the same sign. For a hyperbolic paraboloid, the cross sections in the yz and zx planes are parabolas where one is up-side-down compared to the other, and the cross section in the xy plane is a hyperbola. The hyperbolic paraboloid looks like a saddle.

107

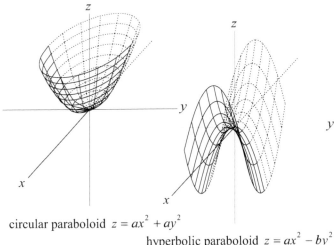

circular paraboloid $z = ax^2 + ay^2$

hyperbolic paraboloid $z = ax^2 - by^2$

A hyperboloid of one sheet looks like a right-circular cylinder, except that it looks hyperbolic rather than rectangular from the side view. The equation for a hyperboloid of one sheet with its axis along the z-axis is $\left(\dfrac{x}{a}\right)^2 + \left(\dfrac{y}{b}\right)^2 - \left(\dfrac{z}{c}\right)^2 = 1$. Its cross sections in the yz and zx planes are hyperbolas, while its cross section parallel to the xy plane is a variable-sized ellipse. A hyperboloid of two sheets can be made with a simple sign change – namely, $\left(\dfrac{x}{a}\right)^2 + \left(\dfrac{y}{b}\right)^2 - \left(\dfrac{z}{c}\right)^2 = -1$ – which splits the hyperboloid into two separate regions.

A quadric surface is the general surface in 3D space for a polynomial equation of the second degree in the coordinates: $ax^2 + by^2 + cz^2 + dxy + eyz + fzx + gx + hy + iz + j = 0$. This includes the sphere, ellipsoid, cylinder, cone, hyperboloid, and paraboloid.

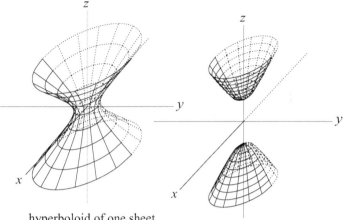

hyperboloid of one sheet

$$\left(\frac{x}{a}\right)^2 + \left(\frac{y}{b}\right)^2 - \left(\frac{z}{c}\right)^2 = 1$$

hyperboloid of two sheets

$$\left(\frac{x}{a}\right)^2 + \left(\frac{y}{b}\right)^2 - \left(\frac{z}{c}\right)^2 = -1$$

A single-holed ring torus in 3D looks like a donut. It can be constructed by taking a right-circular cylinder and bending it into the shape of a circle. The equation for a standard torus with its center at the origin and its ring parallel to the xy plane is

$z^2 = a^2 - \left(b - \sqrt{x^2 + y^2}\right)^2$, where a is the radius of the circular cross

section of the torus and b is the radius of the circular axis of the torus. There are three kinds of standard tori in 3D: The case $a < b$ corresponds to the single-holed ring torus, $a = b$ yields a horn torus, and $a > b$ gives a spindle torus. For the horn torus, the thickness of the torus is just large enough to make the hole vanish (like a donut with no hole); for the spindle torus, the thickness is so large that opposite sides overlap. In general, the torus may have non-circular cross section. The torus may also have multiple holes. The n-torus has n holes.

A curve or a surface may be open or closed. A closed curve bounds a surface, and a closed surface bounds a solid. The interior of a closed curve is a surface area, and the interior of a closed surface is a volume.

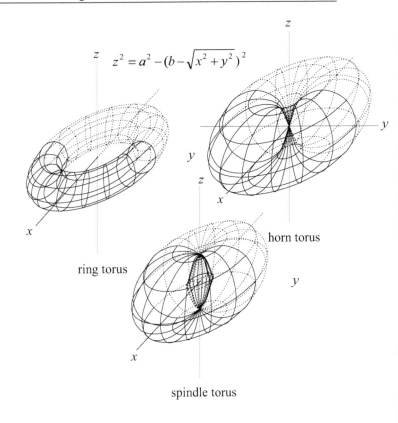

$$z^2 = a^2 - (b - \sqrt{x^2 + y^2})^2$$

horn torus

ring torus

spindle torus

A solid includes the interior of a closed surface in 3D. The ball is a solid corresponding to the interior of a sphere. The sphere is a 2D surface in 3D space, while the ball is the 3D region bounded by the sphere. While the equation for a sphere centered about the origin in 3D is $x^2 + y^2 + z^2 = R^2$, the equation for the corresponding ball is $x^2 + y^2 + z^2 \leq R^2$.

4.2 Hypersurfaces

Although an object moving along a curve experiences just one degree of freedom, a curve can wind its way through higher-dimensional spaces. A circle is a curve that lies in a plane. In 2D, $x^2 + y^2 = R^2$ specifies a circle in the xy plane. In 3D, however, the same equation

$x^2 + y^2 = R^2$ yields a right-circular cylinder. The pair of equations $x^2 + y^2 = R^2$ and $z = 0$, combined together, specify a circle lying in the xy plane. In 3D, two equations are required to specify a curve. A helix is a curve that winds its way through 3D space. The pair of equations $y = \sin z$ and $x = \cos z$ define a helix winding its way around the z-axis. In 4D, the equation $x^2 + y^2 = R^2$ represents the hypersurface of a 4D generalization of a right-circular cylinder, and the pair of equations $x^2 + y^2 = R^2$ and $z = 0$ form a right-circular cylinder along the w-axis. The three equations $x^2 + y^2 = R^2$, $z = 0$, and $w = 0$ are needed to define a circle lying in the xy plane. Three equations are needed to specify a curve in 4D. The two equations $y = \sin az$ and $x = \cos az$ form a surface in 4D: For every point on the usual 3D helix, w is unrestricted. These two equations must be combined with a third equation in order to produce a curve (rather than a surface) in 4D that generalizes the helix.

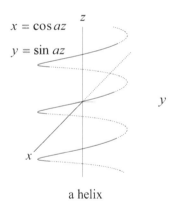

a helix

An object moving along a surface has two degrees of freedom, so that a surface is 2D even if it curves through a higher-dimensional space. In 3D, the equation $x^2 + y^2 + z^2 = R^2$ represents the surface of a sphere centered about the origin. In 4D, however, the same equation $x^2 + y^2 + z^2 = R^2$ is the hypersurface of a generalization of the right-circular cylinder to 4D with spherical cross section. The pair of equations $x^2 + y^2 + z^2 = R^2$ and $w = 0$, combined together, specify the surface of a sphere centered about the origin. In 4D, two equations are needed to specify a 2D surface.

111

In 4D space, a hypersurface is effectively 3D: An object moving in a hypersurface has three degrees of freedom. In 4D, a single equation specifies a hypersurface. A closed hypersurface bounds a hypervolume (aka content) in 4D, just as a closed surface bounds a volume in 3D and a closed curve bounds an area in 2D.

The generalization of a plane to higher dimensions is termed a *hyperplane*. A line is infinite in one dimension, a plane is infinite in two dimensions, and a hyperplane is infinite in three dimensions. The entirety of 3D Euclidean space is a hyperplane. In a 4D space, a 3D space is a subspace, which is why a hyperplane might also be called a 3D subspace. In 4D, there are four orthogonal hyperplanes: xyz, wxy, zwx, and yzw are all mutually orthogonal. The equation $w = 0$ represents the xyz hyperplane, $z = 0$ corresponds to the wxy hyperplane, etc. The general equation for a hyperplane in 4D is $w = ax + by + cz + d$.

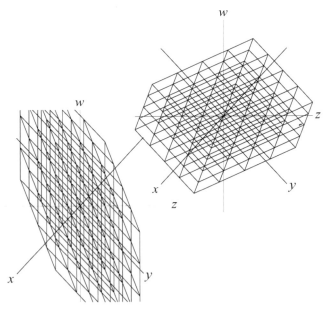

hyperplanes $w = ax + by + cz + d$

In 4D, a hypersolid is 4D. A closed hypersurface bounds a hypersolid, which has hypervolume (aka content). Compare to 3D, where a closed surface bounds a solid, which has volume.

4.3 Anatomy of a Glome

A glome is the 4D generalization of a sphere: It is the locus of points in 4D space equidistant from a common point (i.e. the center). Just as a circle is a 1D curve and a sphere is a 2D surface (since their interiors are excluded), a glome is a 3D hypersurface curving through 4D space. An object moving along the circumference of a circle has one degree of freedom – clockwise/counterclockwise. An object moving along the surface of a sphere has two degrees of freedom – latitude/longitude. An object constrained to move along the hypersurface of a glome experiences 3D motion – it has three degrees of freedom, which can be characterized as latitude, longitude, and hyperlatitude. A circle bounds a 2D area, a sphere bounds a 3D volume, and a glome bounds a 4D hypervolume. In 4D, the equation for a glome with radius R centered about the origin is $x^2 + y^2 + z^2 + w^2 = R^2$.

One way to draw a sphere is to first draw a circle in the plane of the paper – which can be the yz plane – and then, to distinguish the sphere from a circle, draw two more circles – one in the xy plane and one in the zx plane. These three great circles (i.e. a circle whose radius equals that of the sphere) are mutually orthogonal. A glome can be similarly illustrated by drawing six mutually orthogonal great circles in the xy, yz, zw, wx, xz, and yw planes.

A sphere can also be illustrated by drawing circles to represent the two mutually orthogonal directions of motion available to an object constrained to lie on the surface of the sphere – latitude and longitude. A given longitude represents a vertical circle passing through the north and south poles, while a given latitude represents a horizontal circle parallel to the equatorial plane. The circles of fixed longitude are great circles, whereas circles of fixed latitude are smaller (except for the equatorial latitude). Moving north or south is motion along a longitude, while east/west motion occurs along a latitude.

On the hypersurface of a glome, it is possible to move along any combination of three independent directions without getting closer to or further from the center of the glome: In addition to latitude and longitude, the third independent direction could be termed *hyperlatitude*. At any point on the hypersurface of the glome, the circles of latitude, longitude, and hyperlatitude intersect at right angles. To move along a fourth independent direction would require leaving the hypersurface of the hypersphere – i.e. moving inward or outward along the radial coordinate r.

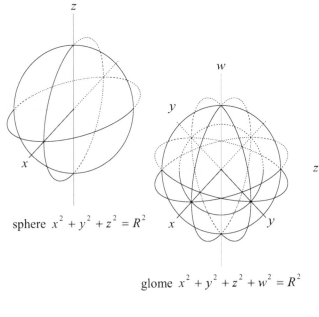

sphere $x^2 + y^2 + z^2 = R^2$

glome $x^2 + y^2 + z^2 + w^2 = R^2$

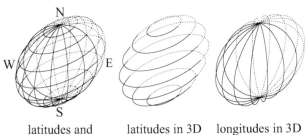

latitudes and   latitudes in 3D   longitudes in 3D
longitudes in 3D

A 2D surface can be drawn on a 3D graph by dividing the surface into 2D tiles. The latitude and longitude circles of a sphere, for example, form a sort of checkerboard pattern on the surface of the sphere. The analogous tiling for a 3D hypersurface drawn on a 4D graph would consist of 3D bricks. (The 3D bricks, like the 2D tiles, would have curved edges, in general – so these are not the usual bricks of buildings that have rectangular sides.) For a glome, the latitude, longitude, and hyperlatitude circles form a 3D checkerboard pattern with 3D bricks. No part of any brick is closer to the center of the glome than any other, nor is any brick closer to the center than any other.

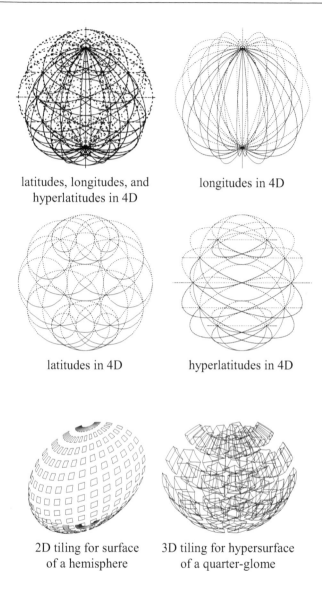

latitudes, longitudes, and
hyperlatitudes in 4D

longitudes in 4D

latitudes in 4D

hyperlatitudes in 4D

2D tiling for surface
of a hemisphere

3D tiling for hypersurface
of a quarter-glome

(As with the tiles of a sphere, no part of any brick is closer to
the center of the glome than any other; nor is any brick closer
to the center than any other.)

Just as a great circle is a circle lying on the surface of a sphere with the same radius as the sphere, a great sphere is a sphere lying on the hypersurface of a glome with radius equal to that of the glome. Analogous to the three mutually orthogonal great circles lying on the surface of a sphere (one for each of the xy, yz, and zx planes), there are four mutually orthogonal great spheres on the hypersurface of a glome (one for each of the xyz, yzw, zwx, and wxy hyperplanes). Two orthogonal great circles intersect at two points (one diameter apart); two orthogonal great spheres similarly intersect at two points (also separated by one diameter).

The 2D cross section of a sphere is a circle (unless the plane is tangent to the sphere, in which case it is a point). The 3D cross section of a glome is a sphere (except for a tangent hyperplane, which similarly touches a singe point on the hypersurface of the glome). A sphere passing through a plane creates a cross section that begins as a point, becomes a growing circle until reaching a great circle, and shrinks back to a point just before vanishing. A glome passing through a hyperplane leaves a similar 3D cross section, except that it is an expanding and contracting sphere.

Slicing a sphere is similar to slicing a tomato; the slices of the sphere have circular cross sections. In the limit that the thickness of the slices approaches zero, the slices become 2D circles. The sphere can be thought of as an infinite number of circles of gradually differing size spaced infinitesimally with their axes aligned. In 2D space, it is not possible to place two circles infinitesimally close to one another (without overlapping): The centers must be separated by the sum of their radii. However, by using the third dimension, the circles can each be rotated 90° and placed infinitesimally close without touching. The set of latitudes are such slices: The latitudes are circles in parallel planes with adjacent circles infinitesimally close.

In 3D space, it is not possible to place two spheres closer together than the sum of their radii (without overlapping); but in 4D space, two spheres could each be rotated 90° (in such a way that is not possible in 3D space) and placed infinitesimally close to one another without touching. Slices of a glome have spherical 3D cross sections. As the thickness of the slices approaches zero, the slices become perfect 3D spheres (i.e. with no thickness in a fourth dimension). The latitudes (and, separately, the hyperlatitudes) of a glome are spheres in parallel hyperplanes with adjacent spheres spaced infinitesimally.

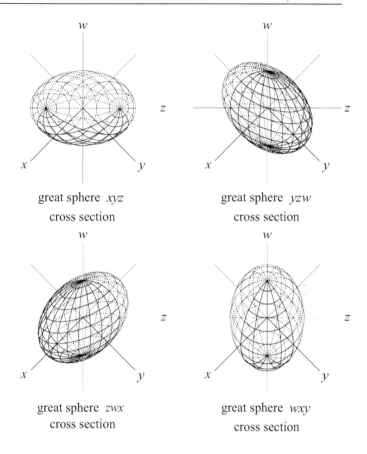

great sphere *xyz*
cross section

great sphere *yzw*
cross section

great sphere *zwx*
cross section

great sphere *wxy*
cross section

(These 3D projections appear as distorted spheres since they are not drawn in perspective.)

Puzzle 4.1: Is there any point to drawing the 3D cross section for a rotating glome – since its cross section is always a sphere?

4.4 Hyperspheres

A hypersphere is a higher-dimensional sphere. The glome is a 4D hypersphere. An N-dimensional hypersphere is actually an $(N-1)$-dimensional hypersurface bounding an N-dimensional hypervolume.

For the discussion and formulas that follow, in this section N will refer to the dimensionality of its hypervolume instead of the the hypersurface.

The N-dimensional sphere is defined as the locus of points in N-dimensional space equidistant from a common point (i.e. the center). The case $N=2$ corresponds to a circle, $N=3$ to a sphere, $N=4$ to a glome, and $N \geq 5$ to a higher-dimensional hypersphere. The case $N=1$ corresponds to the set of points in a 1D space equidistant from a common point, which is simply two points. The 1D "sphere" is given by the equation $x^2 = R^2$, for which the two solutions are $x = \pm R$. The 1D "sphere" is a 0D "surface" bounding a 1D "volume." The terms *sphere*, *surface*, and *volume* – and similarly *hypersphere*, *hypersurface*, and *hypervolume*, in higher dimensions – are used loosely in reference to the general N-dimensional sphere.

An N-dimensional sphere has $\begin{pmatrix} N \\ 2 \end{pmatrix}$ mutually orthogonal great circles. A circle has $\dfrac{2!}{2!0!} = 1$, a sphere has $\dfrac{3!}{2!1!} = 3$, a glome has $\dfrac{4!}{2!2!} = 6$, a 5D hypersphere has $\dfrac{5!}{2!3!} = 10$, and so on. An N-dimensional sphere has $\begin{pmatrix} N \\ 3 \end{pmatrix}$ mutually orthogonal great spheres. A sphere has $\dfrac{3!}{3!0!} = 1$, a glome has $\dfrac{4!}{3!1!} = 4$, a 5D hypersphere has $\dfrac{5!}{3!2!} = 10$, etc. An N-dimensional sphere has $\begin{pmatrix} N \\ M \end{pmatrix}$ mutually orthogonal M-dimensional great subspheres.

	2D	3D	4D	5D	6D	7D	8D	9D
Great Circles	1	3	6	10	15	21	28	35
Great Spheres		1	4	10	20	35	56	84
Great Glomes			1	5	15	35	70	126
Great 5D Spheres				1	6	21	56	126
Great 6D Spheres					1	7	28	84
Great 7D Spheres						1	8	36
Great 8D Spheres							1	9
Great 9D Spheres								1

Puzzle 4.2: What is the difference between the hypersphere pattern and the simplex pattern?

The M-dimensional cross section of an N-dimensional sphere is an M-dimensional sphere (unless the M-dimensional subspace is tangent to the N-dimensional sphere, in which case it is a point). The 2D cross section of any $N \geq 3$-dimensional sphere is a circle, the 3D cross section of any $N \geq 4$-dimensional sphere is a sphere, etc.

4.5 Quadric Hypersurfaces

The equation for a quadric hypersurface has at least one coordinate (i.e. x, y, z, w, etc.) squared; there may also be cross terms (e.g. $3xy$ or $2wz$) and a constant. Quadric hypersurfaces are the 4D generalizations of the quadric surfaces.

A hyperellipsoid is a higher-dimensional generalization of the ellipse. The 3D cross sections in the xyz, yzw, zwx, and wxy hyperplanes are ellipsoids. The equation for a 4D hyperellipsoid with its symmetry axes along the coordinate axes is

$$\left(\frac{x}{a}\right)^2 + \left(\frac{y}{b}\right)^2 + \left(\frac{z}{c}\right)^2 + \left(\frac{w}{d}\right)^2 = 1.$$

The ellipsoid, spheroid, and the sphere are distinguished by their projections onto the xy, yz, and zx planes. With the symmetry axes along the coordinate axes, the three projections are: ellipses of differing eccentricity for an ellipsoid; a circle and two identical ellipses for a spheroid; and identical circles for a sphere. The equation for such an ellipsoid is $\left(\frac{x}{a}\right)^2 + \left(\frac{y}{b}\right)^2 + \left(\frac{z}{c}\right)^2 = 1$, where the constants a, b, and c differ. If two of the constants are equal, e.g. $b = c$, the equation instead describes a spheroid. The third constant, in this case a, determines whether it is a prolate ($|a| > |b|$) or oblate ($|a| < |b|$) spheroid. If all three constants are equal, i.e. $a = b = c$, the equation describes a sphere.

The equation for a 4D hyperellipsoid with its symmetry axes along the coordinate axes is $\left(\frac{x}{a}\right)^2 + \left(\frac{y}{b}\right)^2 + \left(\frac{z}{c}\right)^2 + \left(\frac{w}{d}\right)^2 = 1$, where the four constants differ. If all of the constants are equal, the equation describes a glome. If some, but not all, of the constants are equal, the equation describes a hyperspheroid. Illustrations of hyperellipsoids and hyperspheroids look similar to hyperspheres, except for stretching/contracting one or more axes.

A hyperspheroid is a higher-dimensional generalization of the spheroid. While there are two distinct types of spheroids – prolate and oblate – there are five types of 4D hyperspheroids. If three constants are equal, e.g. $b = c = d$, the hyperspheroid can be prolate $(|a| > |b|)$ or oblate $(|a| < |b|)$; if only two constants are equal, e.g. $c = d$, there are three distinct kinds of hyperspheroids, depending on whether $|c|$ is smaller than both $|a|$ and $|b|$, $|c|$ is greater than both $|a|$ and $|b|$, or $|c|$ is greater than one and less than the other of $|a|$ and $|b|$.

Puzzle 4.3: How many distinct types of 5D hyperspheroids are there?

A hypercylinder is a generalization of the cylinder to higher dimensions. There are two ways to generalize the definition of a 3D cylinder to 4D: A cubinder is the locus of points in 4D equidistant from an axis, whereas a spherinder is the locus of points equidistant from a plane.

A spherinder with its axis along the w-axis is specified by the equation $x^2 + y^2 + z^2 = R^2$. The x-, y-, and z- coordinates form a sphere about any point on the w-axis. Just as a right-circular cylinder can be thought of as a series of parallel circles with a common axis, a spherinder can be thought of as a series of parallel spheres with a common axis, where the centers of adjacent spheres are infinitesimally close. The spherinder determined by the equation $x^2 + y^2 + z^2 = R^2$ (which is a sphere in 3D, but a spherinder in 4D) has the cross section of a sphere in the xyz hyperplane and a right-circular cylinder in the yzw, zwx, and wxy hyperplanes.

A cubinder with its symmetry plane the zw plane is specified by the equation $x^2 + y^2 = R^2$. The x- and y-coordinates form a circle about any point in the zw plane. The cubinder determined by the equation $x^2 + y^2 = R^2$ has the cross section of a 3D right-circular cylinder in the xyz and wxy hyperplanes and a pair of parallel planes in the zwx and yzw hyperplanes.

Puzzle 4.4: How many ways are there to generalize a cylinder to 5D?

An N-dimensional spherinder is the locus of points in N-dimensional space equidistant from a line (the axis of the spherinder). The pair of lines $x = \pm R$, which are equidistant from the y-axis, form a 2D spherinder; a cylinder is a 3D spherinder; a spherinder, by default, is 4D; a 5D spherinder might better be termed a *glominder*; etc. The glominder determined by the equation $x^2 + y^2 + z^2 + w^2 = R^2$ (this is a glominder in 5D, but a glome in 4D) has the cross section of a glome in the $xyzw$ 4D hyperplane and a spherinder in the $yzwv$, $zwvx$, $wvxy$, and $vxyz$ 4D hyperplanes.

An N-dimensional cubinder is the locus of points in N-dimensional space equidistant from a plane. The two parallel planes $x = \pm R$, which are equidistant from the yz plane, form a 3D cubinder. The 5D cubinder determined by the equation $x^2 + y^2 + z^2 = R^2$ (which is a sphere in 3D or a spherinder in 4D, but a cubinder in 5D) has the cross section of a sphere in the xyz 3D hyperplane; a cylinder (a 3D spherinder) in the xyw, xyv, xzw, xzv, yzw, and yzv 3D hyperplanes; a pair of parallel planes (a 3D cubinder) in the xwv, ywv, and zwv 3D hyperplanes; a spherinder in the $xyzw$ and $xyzv$ 4D hyperplanes; and a cubinder in the $xywv$, $xzwv$, and $yzwv$ 4D hyperplanes.

In higher dimensions, it is possible to form even more varieties of hypercylinders. For example, in 5D, in addition to 5D spherinders and 5D cubinders, there is a 5D hypercylinder which is the locus of points in 5D space equidistant from a hyperplane. In 4D, this would be a pair of parallel hyperplanes, but in 5D, the 2D cross section perpendicular to the hyperplane symmetry axis is a circle. In 6D, there is a 6D spherinder with a cross section of a 5D hypersphere, a 6D cubinder with a cross section of a glome, a hypercylinder with a cross section of a sphere, and another hypercylinder with a cross section of a circle (where these cross sections are in the hyperplane perpendicular to the axis of the hypercylinder). These cross sections can be changed to produce generalizations of right-elliptical cylinders, right-hyperbolic cylinders, and so on. Also, the cylinders may be slanted (i.e. not *right*).

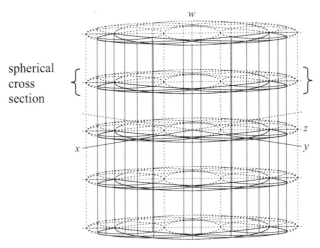

spherical
cross
section

Although the cross section of the spherinder in the *xyz* hyperplane is a sphere, it looks flat in the diagram due to projections onto the plane of the page. The 3D space of the spherical cross section is perpendicular to the axis of the hypercylinder.

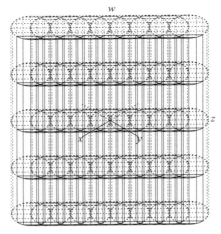

Every point in the *zw* plane lies in the center of a circle in the *xy* plane. None of these circles intersects the *zw* plane nor do any of these circles intersect one another, although the projected image creates these illusions.

The difference between a right-circular cone and a right-circular cylinder is that the circular cross section of the cone varies in size whereas the circular cross section of the cylinder is constant. For every hypercylinder, there is an analogous hypercone – the higher-dimensional generalization of a cone. There is a 4D hypercone similar to the spherinder, except that the radius of the spherical cross section varies linearly along the axis, and there is a 4D hypercone analogous to the cubinder, where the difference is that the circular cross section varies linearly with the height of the hypercone.

The equation for an ellipse with its symmetry axes along the coordinate axes, $\left(\dfrac{x}{a}\right)^2 + \left(\dfrac{y}{b}\right)^2 = 1$, becomes a hyperbola with a simple sign change, $\left(\dfrac{x}{a}\right)^2 - \left(\dfrac{y}{b}\right)^2 = 1$. The ellipse becomes an ellipsoid by adding a similar term for the third independent coordinate: $\left(\dfrac{x}{a}\right)^2 + \left(\dfrac{y}{b}\right)^2 + \left(\dfrac{z}{c}\right)^2 = 1$. There are two distinct types of hyperboloids: A hyperboloid of one sheet is obtained by one sign change, e.g. $\left(\dfrac{x}{a}\right)^2 + \left(\dfrac{y}{b}\right)^2 - \left(\dfrac{z}{c}\right)^2 = 1$, while a hyperboloid of two sheets is obtained via two sign changes, e.g. $-\left(\dfrac{x}{a}\right)^2 - \left(\dfrac{y}{b}\right)^2 + \left(\dfrac{z}{c}\right)^2 = 1$. (Note that this latter equation is equivalent to $\left(\dfrac{x}{a}\right)^2 + \left(\dfrac{y}{b}\right)^2 = \left(\dfrac{z}{c}\right)^2 - 1$.) More sign changes are possible in higher dimensions.

A hyperhyperboloid is the higher-dimensional generalization of the hyperboloid. For symmetry about the coordinate axes, the equation for a hyperboloid can involve one, two, or three sign changes compared to the equation for a hyperellipsoid: The equation $\left(\dfrac{x}{a}\right)^2 + \left(\dfrac{y}{b}\right)^2 + \left(\dfrac{z}{c}\right)^2 - \left(\dfrac{w}{d}\right)^2 = 1$ describes a hyperhyperboloid of one hypersheet, the equation $\left(\dfrac{x}{a}\right)^2 + \left(\dfrac{y}{b}\right)^2 - \left(\dfrac{z}{c}\right)^2 - \left(\dfrac{w}{d}\right)^2 = 1$ represents a different type of hyperhyperboloid of one hypersheet, and the equation $\left(\dfrac{x}{a}\right)^2 + \left(\dfrac{y}{b}\right)^2 + \left(\dfrac{z}{c}\right)^2 - \left(\dfrac{w}{d}\right)^2 = -1$ specifies a hyperhyperboloid of two hypersheets.

123

spherical
cross
section

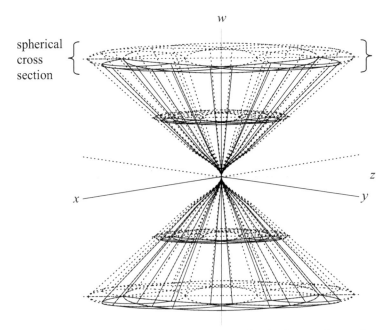

a hypercone with spherical cross section in the *xyz* hyperplane

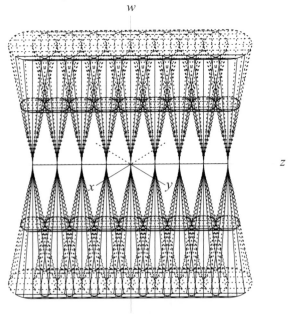

a hypercone with circular cross section in the *xy* plane

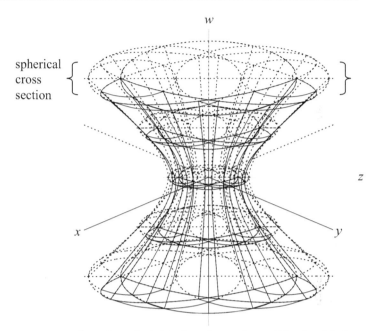

a hyperhyperboloid of one hypersheet of the first kind

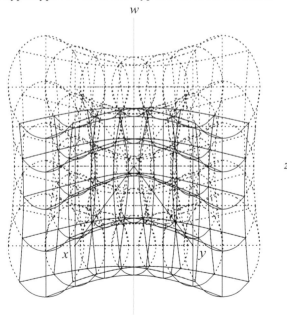

a hyperhyperboloid of one hypersheet of the second kind

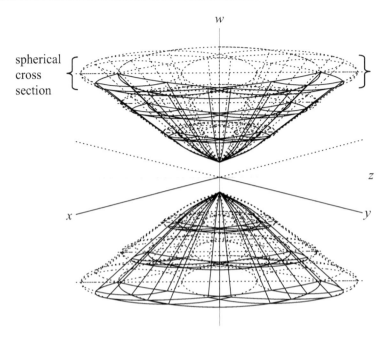

spherical
cross
section

a hyperhyperboloid of two hypersheets

A hyperparaboloid is a higher-dimensional paraboloid. As in 3D, there are two kinds of 4D hyperparaboloids: a spherical (or, more generally, ellipsoidal) hyperparaboloid and a hyperbolic hyperparaboloid. The equation for a spherical hyperparaboloid with its vertex at the origin and symmetry axis along the positive w-axis is $w = a(x^2 + y^2 + z^2)$. The cross section can be made ellipsoidal by giving different coefficients to x^2, y^2, and z^2. The equation for a hyperbolic hyperparaboloid is obtained by introducing a minus sign, e.g. $w = \left(\dfrac{x}{a}\right)^2 + \left(\dfrac{y}{b}\right)^2 - \left(\dfrac{z}{c}\right)^2$. The spherical hyperparaboloid is shaped like a 4D bowl, and the hyperbolic hyperparaboloid is shaped like a 4D saddle.

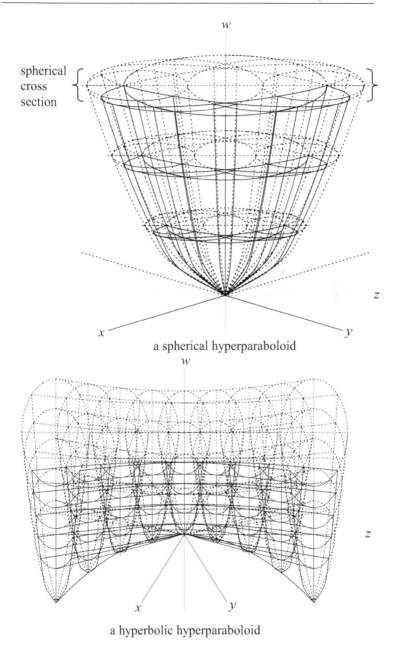

spherical cross section

a spherical hyperparaboloid

a hyperbolic hyperparaboloid

4.6 Hypertori

One way to make a single-holed ring torus is to bend a finite right-circular cylinder into a donut. Whereas the axis of a right-circular cylinder is a line, the axis of a single-holed ring torus is a circle. For either, a cross section perpendicular to the axis is a circle.

A hypertorus is a higher-dimensional torus. There are two ways to generalize a single-holed ring torus to 4D. One way is to bend a finite spherinder into a circle, and another way is to bend a finite cubinder into a sphere. The former has spherical cross section perpendicular to the circular axis, and the latter has circular cross section perpendicular to the bisecting sphere (in this case, the "axis" of 3D has generalized to a surface). The circular 4D hypertorus generalizes the 3D ring torus by increasing the dimensionality of the cross section, preserving the structure of the ring, whereas the spherical 4D hypertorus increases the dimensionality of the ring structure, preserving the cross section. A ring has inner and outer circles, and the bisecting circle defines the axis of the ring. The hyperring structure of the spherical 4D hypertorus has inner and outer spheres; the bisector of these is also a sphere (not an axis).

A single-holed circular 4D hypertorus is a ring with spherical 3D cross section. Every point on the circular axis of the ring lies at the center of a sphere, where the three independent directions of the space within the sphere are all perpendicular to the circular axis of the ring. If the ring lies in the xy plane with its center at the origin, the equation for the single-holed circular 4D hypertorus is

$$z^2 + w^2 + \left(b - \sqrt{x^2 + y^2}\right)^2 = a^2,$$ where b is the radius of the circular

axis in the xy plane and a is the radius of the spherical cross section in the yzw or zwx hyperplanes. The three independent directions of the spherical cross sections are: radially inward/outward in the xy plane, up/down along the $\pm z$-direction, and ana/kata along the $\pm w$-direction; the fourth independent direction, perpendicular to these, is tangential to the circular axis in the xy plane.

While the circular 4D hypertorus can be thought of as a circular axis where every point on the circular axis is the center of a sphere (where none of these spheres overlap), the spherical 4D hypertorus is a spherical bisector where every point on the spherical bisector is the center of a circle (where none of these circles overlap). The two independent directions of the circular 2D cross sections are perpendicular to the two independent directions of the spherical bisector. The equation for a spherical 4D hypertorus with the spherical bisector lying in the xyz hyperplane and centered about the origin is

$$w^2 = a^2 - \left(b - \sqrt{x^2 + y^2 + z^2} \right)^2$$, where a is the radius of the circular

cross section in the xw, wy, or zw planes and b is the radius of the spherical bisector in the xyz hyperplane. The two independent directions of the circular cross sections are radially inward/outward in the xyz hyperplane and ana/kata in the $\pm w$-directions, while the two independent directions of the spherical bisector are tangential to the spherical bisector (latitudes and longitudes).

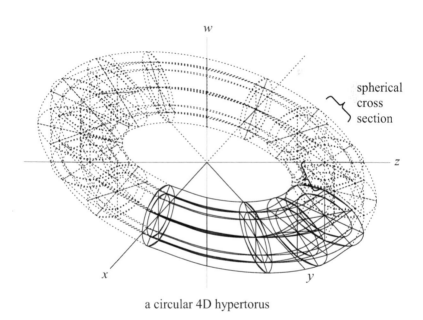

a circular 4D hypertorus

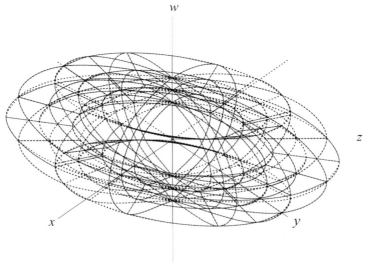

a spherical 4D hypertorus

5 A Hypothetical Hyperuniverse

Now we will imagine a 4D universe where the fourth dimension is just like the three known dimensions. Of course, any extra dimensions in the known universe must be hidden in some way, e.g. via compactification, yet the 'toy' model considered here will serve some useful purposes. For one, it will give us a chance to apply some of the geometric concepts and visualization skills that we have learned. We will also more fully grasp the significance of the extra degree of freedom available in 4D. This extra freedom is there at the particle level even if the extra dimension is compact, so it is worth studying this simpler noncompact case first. Aside from this, a tour through the fourth dimension can be a fascinating journey, practical or not.

5.0 Hyperuniverse NCF4D-U

The most straightforward way to generalize the three observed spatial dimensions of the known universe to four spatial dimensions is to add a fourth dimension as follows: The fourth dimension is assumed to be noncompact as well as flat (i.e. not curved, except in the spacetime sense of general relativity), and all objects in this hyperuniverse are assumed to be able to experience the full 4D freedom of motion. Such a hyperuniverse might be abbreviated NCF4D-U, to distinguish it from other possible hyperuniverses (NC means noncompact, F is flat, 4D refers to the dimensionality of the space, and U is for universal – meaning that all of the particles of the hyperuniverse can propagate into all four dimensions). This is not a standard notation, but it will serve as a useful abbreviation for our purposes.

Common human experience suggests that the known universe can not be quite like the NCF4D-U hyperuniverse – macroscopic objects, at least, evidently are unable to move in four independent directions. If the known universe is a hyperuniverse, the extra dimensions must either be curled up with such a microscopic size that they have not yet been detected or the known particles of the known universe must not have full freedom in the extra dimensions – or both. Although the NCF4D-U hyperuniverse is not a plausible model for the known universe, there is some motivation for considering it in detail.

Here are some reasons for studying the NCF4D-U hyperuniverse:

- It may help to provide a better overall understanding of higher-dimensional space.
- There are useful mathematical and physical relationships with independent variables similar to higher-dimensional space, which may have useful analogies with NCF4D-U.
- As the most straightforward generalization of 3D space, it is a useful prelude to considering compact extra dimensions.
- It provides insight into higher-dimensional motion, which may occur at least at the microscopic level in the known universe.
- Even if not plausible, it may be interesting and fun.

5.1 The Physics of NCF4D-U

We will consider a hyperuniverse where the laws of physics are straightforward generalizations of the laws observed in the known universe. It is certainly easy to imagine a hyperuniverse with different physical laws. However, one motivation for studying a NCF4D-U universe is that it will serve as a useful prelude to studying compact extra dimensions that may exist in the known universe. To this end, we need the physical laws to agree with those observed in the known universe in appropriate limits – namely, in the limit that the extra dimension is removed.

Newton's law of universal gravitation, which has a major consequence on macroscopic motion in the known universe, and Coulomb's law of electrical attraction and repulsion are observed to be inverse-square laws in the known universe. Each of these laws follows from Gauss's law, which relates the flux of field lines (gravitational in one case, and electric in the other) to the source (mass being the source of gravitational field lines, and charge the source of electric field lines). In higher-dimensions, Gauss's law requires that these inverse-square laws become inverse-cube (or greater power) laws. This follows from the extra degrees of freedom available in a higher-dimensional space, which has a marked effect on the flux of field lines. We will examine this in greater detail in Chapter 7. For now, it is important to realize that:

Thus, some laws of physics must be different in the hyperuniverse: Either Gauss's law must not hold, or the usual inverse-square laws must change power; preserving Gauss's law is the much more compelling option for developing a sound theory analogous to the known universe, and is consistent with superstring theory – our primary inspiration for considering extra dimensions. Henceforth, it will be assumed that the laws of physics in NCF4D-U are a straightforward generalization of the laws of physics of the known universe; in particular, Gauss's law is preserved.

The power in Newton's law of universal gravitation affects the allowed orbits. In NCF3D-U (which *is* consistent with the observed universe), the inverse-square law of gravity allows the following two-body orbits: a line, a circle, an ellipse, a parabola, and a hyperbola. In higher dimensions, a different power law comes with different orbits.

Gauss's law requires that gravity follow an inverse-cube law in NCF4D-U. Celestial objects, such as stars, planets, and moons, would be hyperspherical in shape. Such planets would spin on an a plane (not a linear axis, like the earth). These hyperspheroids would bulge at the equatorial hyperplane – similar to the earth's bulge at its equatorial plane. The orbits of celestial objects, such as moons, planets, and comets, would be altered to some extent. Although gravity would be an inverse-cube law rather than an inverse-square law, it would still be approximately constant for small changes in altitude (relative to the radius of the planet). Thus, the local experiences that planetary inhabitants would have with gravity would not be noticeably different inasmuch as gravitational acceleration would be approximately constant near the hypersurface of the planet. The behavior of a falling apple would be very similar to an apple falling near the earth, as would the motion of projectiles (unless they were very long-range or high-altitude, in which case the mathematical differences in the underlying ballistics would be significant).

There would be a similar change in the inverse-square law of Coulomb's law to an inverse-cube law, which would affect the mathematical form of the interactions of electric charges, which may lead to important differences in physical phenomena that involve these interactions. However, some basic qualitative interactions, such as the attraction of opposite charges and repulsion of like charges, would be unchanged. Basic atomic structure would be modified by a fourth noncompact, universal extra dimension. Orbital and spin angular momentum, which, along with energy, affect the electronic structure of an atom, would have more components. Thus, the extra degree of freedom may lead to a much richer periodic table of elements, which could have significant impacts on chemistry in higher dimensions. This, in turn, affects the biological and other sciences. Nonetheless, it is still possible to conceive of life in NCF4D-U with some similarities to life in the known universe: with elements bonding together to form compounds and molecules that make up gases, liquids, and solids; with stars, planets, moons, and comets; and with 4D animals and even 4D humanoid creatures. There would be some significant contrasts, too, perhaps as different as life in an analogous 2D universe would be.

We will examine higher-dimensional physical laws in more detail in Chapter 7. In the meantime, an important point to bear in mind is that gravity near the surface of a planet would be rather similar to gravity near earth's surface. The more significant difference is that there would be three, rather than two, independent directions to move along for which height would not change – forward/backward, left/right, and ana/kata.

5.2 Humanoids in the Fourth Dimension

Let us imagine a human-like inhabitant of NCF4D-U. Such a being would have a torso, a head, and arms and legs, but we will see reasons that many features would be as different from a human as a 2D humanoid would be.

The skin of a 4D humanoid would be the hypersurface that bounds the hypervolume of the body. The skin of a human being is effectively 2D: A bug crawling on the skin has two degrees of freedom. The skin of a 4D humanoid, however, would be effectively 3D: A 4D bug crawling on the skin of a 4D humanoid would experience three degrees of freedom. The skull of a 4D would be somewhat glome-shaped, but no more than the skull of a human is spherical.

A 4D humanoid would sense via the four independent directions of 4D space. Forward/backward would be defined by the direction in which the humanoid is moving, which provides a sense of depth. Upward/downward would be defined by the gravitational pull of the planet on which the 4D humanoids are inhabitants, which provides a sense of height (or altitude). In 3D, left/right is the only direction perpendicular to up/down and front/back, but in 4D, there is also ana/kata. From the perspective of a 4D humanoid, the ana/kata direction would be more like left/right than like front/back or up/down. Thus, there would be two kinds of width – there would be one width corresponding to left/right, and a hyperwidth for ana/kata.

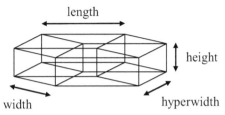

the 4 Cartesian dimensions of 4D

In 4D, a pupil could experience three degrees of freedom on the hypersurface of an eyeball: The pupil could move up and down to gauge height, left and right to gauge width, and ana and kata to gauge a hyperwidth; multiple eyes would help to gauge depth. A third eye could be a significant advantage in gauging depth in 4D. The plane of the triangle made by connecting the three eyes would lie in the left/right-ana/kata plane. Peripheral vision would be best if it could detect motion at the extreme ana and kata directions of vision as well as the extreme left and right directions.

It would be an advantage for a humanoid to have four ears – one on each of the left, right, ana, and kata sides of the head, which correspond to the directions perpendicular to front/back and up/down.

In 3D, a human is among the few bipeds. Many animals walk on four legs; walking on two legs requires better balance. In 2D, it would be very easy to stand on two legs, while a one-legged being would need good balance. In 3D, only three legs are needed, in principle, for easy stability, since any three nonlinear points (i.e. the feet) lie in a plane, but since animal limbs tend to come in pairs, the choice is really between two or four. Two points define a line, which is why a 2D being with two legs would balance with ease.

Four nonplanar points define a hyperplane. Since a 4D floor would be a hyperplane, four legs are enough for easy balance, in principle; however, expecting pairs of legs to tend to be roughly parallel, by analogy with the known universe, six legs may be more common for the same reason that walking on three legs in 3D would be somewhat awkward. A 4D humanoid with two legs would require a better sense of balance than a human, otherwise it would be clumsier. A 4D humanoid might be better off with four legs – one left, right, ana, and kata leg – in which case balance would be very easy to develop compared to humans in 3D.

A human can throw a ball overhand or sidearm. In 4D, there would be two types of sidearm throws – one for left/right and the other for ana/kata. For maximum arm mobility, it would be handy for a 4D humanoid to have four arms – one on the left, right, ana, and kata sides.

135

In 3D, an extended arm is roughly 1D; bending the elbow, the arm is 2D; twisting at the elbow and/or shoulder provide 3D mobility. In 4D, two elbows on each arm would allow maximum mobility with each arm if the two elbows on each arm bent in perpendicular directions. There would also be two independent types of twists possible for each lower arm, middle arm, and upper arm; more advanced joints and sockets would be needed to achieve this additional twisting.

Grasping in 4D would be easier with two kinds of opposable thumbs, where the fingers, first thumb, and second thumb can bend in three mutually perpendicular directions. While there are two types of hands in 3D (left and right), there would be four types of hands possible in 4D with two different kinds of opposable thumbs (described geometrically in Section 5.3) – the same number of arms that would be useful in 4D. An improvement over two opposable thumbs would be one thumb and two sets of fingers that can bend in mutually orthogonal directions.

A human can rotate the torso and thighs left or right through a simple rotation of the hips. It would be ideal for a 4D humanoid to be able to rotate hips in two independent directions – left/right or ana/kata, or any combination of these. The hips would also need to allow for bending forward, just as in 3D.

A human can look to the left or right by rotating the neck. A 4D humanoid would benefit from being able to rotate the neck left/right or ana/kata, or any combination of these two independent directions.

The surface of a human eyeball has two degrees of freedom. The image that forms in the mind, based on the image that forms in the eyes, is 2D – similar to a photograph or a picture painted on a canvas. One notable difference is that the eyes and head can move, bringing with it changes in the way the universe is perceived; whereas a photograph or painting looks the same from any angle. However, what is left or right or what is above or below is more obvious, and the perception of depth is more subtle, in principle – depth is gauged largely from experience. The eyeball of a 4D humanoid would have a hypersurface with three degrees of freedom. This would result in 3D images where left/right, high/low, and ana/kata are obvious, but where depth is more difficult to gauge.

The upper and lower arms, upper and lower legs, neck, fingers and toes, and torso of humans are somewhat cylindrical – not perfect right-circular cylinders, obviously, but much closer to finite cylinders than to the other basic geometries. Since a cylinder can be generalized to a cubinder or spherinder in 4D (described in detail in Chapter 4), there are vastly different shapes possible for a 4D humanoid.

In 4D, clothing would be manufactured with two different waist sizes: There would also be a hyperwaist measurement because there would be two independent ways to wrap a tape measure around the midsection of a 4D humanoid. In addition to shoe size, which corresponds to the length of a foot, there would be width and hyperwidth.

5.3 Higher-Dimensional Chirality

Let us return to the issue of handedness in 4D, and consider it in more detail. In 3D, humans have two types of hands – left and right – which look like mirror images of one another. The two hands are distinctly different in the sense that there is no way to turn a left hand into a right hand (or vice-versa) by rotating it in 3D space (where a reflection does not count as a rotation). In 3D, hands come in one of two chiral states – left-handed or right-handed. Chirality is a mathematical term used to describe an asymmetric object that can't be superimposed on its mirror image, such as a right hand.

Other objects in the known universe have a similar sense of chirality. For example, a 3D coordinate system can be right-handed or left-handed. Pointing the fingers along x, then curling them along y (without rotating the hand), the thumb (in the opposed position) will point along z for only one of the two hands. The term even extends to physics, where electrons have two different spin states available, and chemistry, where molecules may have an asymmetrical form. Generalizing chirality to N dimensions, there can be more than two chiral states.

Hands are not the only objects that possess a sense of chirality. A simple configuration of blocks can illustrate chirality. This is especially useful in higher dimensions, since it is much easier to visualize a hypercube than a hyperhand.

In 2D, there would be two types of hands – L-shaped and ⅃-shaped – where one leg of the L is longer than the other, since fingers are longer than thumbs. These L and ⅃ shapes would not correspond to left and right hands because it is not possible for a 2D humanoid to have arms to the left or right – these must either be front and back hands (attached to front and back arms) or top and bottom hands (two arms in front, one above the other).

fingers extended along x

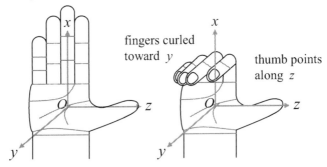

a right-handed coordinate system

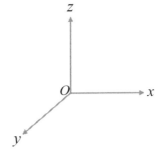

Puzzle 5.1: Is this coordinate system left- or right-handed?

A configuration of four squares can illustrate 2D chirality. There are six distinct configurations of four squares where adjacent squares join at an edge: all in a row, an L shape, a J shape, a T shape, a large square, and a zigzag. The L and J shapes are analogous to the two types of 2D hands. The L and J shapes are distinct in 2D – it is not possible to rotate an L-shaped hand in 2D and turn it into a J-shaped hand (recalling that one leg of the L is longer than the other, this would require a reflection, not a rotation, in 2D). In three or more dimensions, however, the 2D L and J shapes do not count as two distinct shapes: It is possible to rotate an L-shaped hand through the third dimension so that it becomes a J-shaped hand. If you rotate an L in a plane, it will never look like a J. However, if you pick the L up out of the plane – into the third dimension – then flip it over and return it to the plane, it will look like a J.

In 3D, there are eight distinct configurations of four cubes where pairs of cubes join at a face. Three of these configurations have cubes extending in all three dimensions. Two correspond to left and right hands (with fingers bent forward into an L-shape and thumb opposed). While there are two types of hands in 2D and 3D, a fundamental difference is that in 3D the thumb can bend in multiple directions.

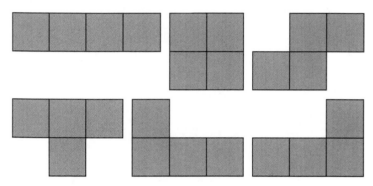

six distinct types of 2D block configurations consisting of four squares

There are 31 distinct 4D configurations consisting of five tesseracts where adjacent tesseracts share bounding cubes. These include 1 linear configuration, 11 planar configurations, 12 configurations that extend in three independent directions, and 7 fully 4D configurations. Four of the fully 4D configurations correspond to possible 4D hands that are analogous to the 3D hands of humans.

A 4D hand might have one set of fingers and two opposable thumbs, for which there are four distinct types of hands; or it might have one thumb and two sets of fingers that bend in mutually orthogonal directions, for which there are still four distinct types of hands. The usual left and right hands of 3D are not distinct in 4D because it is possible to rotate a left 3D hand in such a way as to turn it into a right 3D hand in 4D (of course, such a rotation is not possible in 3D). This is analogous to L- and ⌐-shaped hands, which are distinct in 2D, but not in 3D because there is a rotation possible in 3D, which is not available in the plane of the L- and ⌐-shaped hands, that effectively causes the L- or ⌐-shaped hand to reflect.

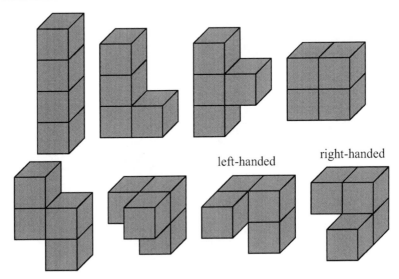

left-handed   right-handed

eight distinct types of 3D block configurations consisting of four cubes

An N-dimensional object can be reflected via a 180° rotation in $(N+1)$-dimensional (or higher) space, even when such a rotation could not be achieved through rotation within the N-dimensional space. For example, consider the L-shape formed by two line segments $(0,0) - (0,2)$ and $(0,0) - (1,0)$. Within the 2D space of the object – i.e. the xy plane – the L-shaped object can rotate about any point in the plane. The corresponding rotation in 3D space is to rotate about an axis perpendicular to the plane. No such rotation can change the L shape into a Γ shape. However, in 3D space, more types of rotation are possible – e.g. rotation about the x- or y-axis (or any linear combination of rotations about the three coordinate axes). Upon a rotation about the y-axis, the endpoints of the two line segments become $(0,0) - (0,2)$ and $(0,0) - (-1,0)$, which is a Γ shape. This 3D rotation is equivalent to a 2D reflection. Similarly, a 3D left hand can be rotated into a 3D right hand in 4D space.

5.4 Construction in the Fourth Dimension

Now we will consider what types of objects such 4D humanoids might build. First, we will look at simple 4D machines, which would serve as the basic building blocks upon which all machinery would be built. Then we will examine possible 4D structures, such as buildings.

It is reasonably straightforward to generalize 3D simple machines to 4D, but it is more challenging to try to imagine 4D simple machines that do not have a direct analogy in 3D – just as it would be very difficult for a 2D being to invent a screw or visualize a wheel turning on an axle. We will focus on the direct analogies, but with some ingenuity, you may be able to invent a simple 4D machine that does not have a direct analog in 3D.

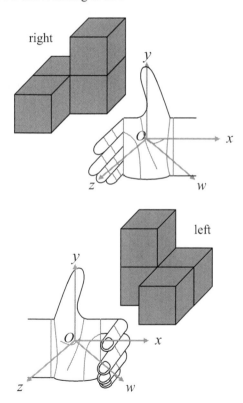

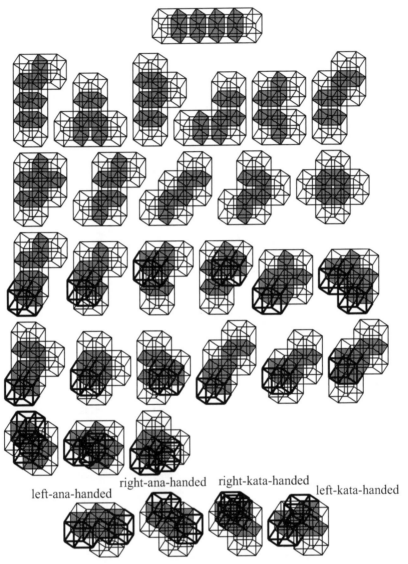

left-ana-handed   right-ana-handed   right-kata-handed   left-kata-handed

31 distinct types of 4D block configurations consisting of five tesseracts

A typical 3D wheel has a disc-like structure, which can be thought of as a circle with a small thickness in the third dimension or a finite right-circular cylinder that is very short compared to its diameter. A hole in the center of a 3D wheel allows it to be slipped over an axle, which is rod-shaped. The axle makes it possible to connect the wheel as part of a complex machine, such as a bicycle or a car. An axle perpendicular to the plane of the wheel is possible in 3D, but not in 2D.

There are two basic ways to generalize a wheel to 4D: a short, finite spherinder or a short, narrow, finite cubinder. The 4D wheel with spherinder shape – a hypercylinder with spherical 3D cross section – would be like a sphere with a small thickness in the fourth dimension. The cubinder-shaped wheel would have circular 2D cross section, but two independent kinds of thickness.

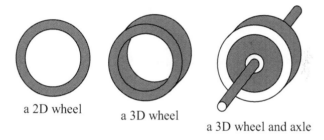

a 2D wheel   a 3D wheel

a 3D wheel and axle

The axle of a spherinder-shaped 4D wheel would pass through perpendicularly to the sphere just as an axle passes perpendicularly through the circle of a 3D wheel. Just as a typical 3D wheel is a large, short cylinder with an axle that is a long, thin cylindrical rod, a spherinder-shaped 4D wheel would be a large, short spherinder with an axle that is a long, thin spherinder. The axis of the spherinder-shaped rod would be perpendicular to the three independent directions of the spherical cross section of the wheel. While there is only one direction perpendicular to the spherical 3D cross section of a spherinder-shaped 4D wheel, there is a whole plane perpendicular to the circular 2D cross section of a cubinder-shaped 4D wheel. Hence, the axle of a cubinder-shaped 4D wheel would not extend mostly along a rod, as in 3D, but along a plane. In this case, the axle would be a cubinder with long length, but shorter width, where both the length and width are long compared to the circular 2D cross section, while the length and width would reverse roles for the wheel.

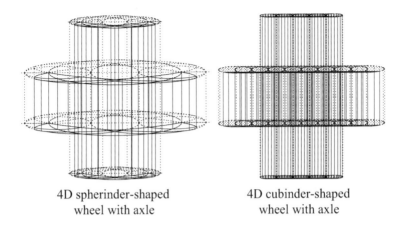

4D spherinder-shaped
wheel with axle

4D cubinder-shaped
wheel with axle

A ramp would be an inclined hyperplane rather than an inclined plane in 4D; an inclined plane in 4D would be as inadequate as an inclined line is in 3D.

A 3D pulley is a wheel and axle that suspends from a ceiling or clamps to a support rod, which features a groove in the rim of the wheel for a cord to wrap partway around it. A basic function of a pulley is effectively to change the direction of an applied force. For example, a cord pulling a box horizontally can be passed over a pulley and tied to a weight, in which case the gravitational force exerted on the weight does work not only to move the weight but the box as well. A 4D pulley would be similar, except that the 4D wheel and axle would be spherinder or cubinder in shape. The groove could be designed for a long, thin cord, as in 3D, or for a long, wide, thin sheet. A sheet could drape over a spherinder-shaped pulley like a sheet drapes over a kid to make a ghost costume for Halloween – but without the necessity of cutting any holes in the sheet to connect the pulley's supports!

The basic function of a lever in 4D should not be expected to change much from 3D, since the common use of a lever in 3D is no different than it would be in 2D. Either way, an effectively 1D arm is balanced on a fulcrum. The principal behind the lever is that a longer lever arm requires less force to balance the torque. This allows, for example, a heavy load to be lifted more easily. The lever also provides a means – namely, a balance – for measuring weight.

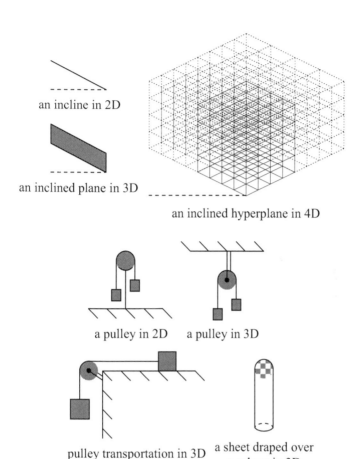

an incline in 2D

an inclined plane in 3D

an inclined hyperplane in 4D

a pulley in 2D   a pulley in 3D

pulley transportation in 3D   a sheet draped over a sphere in 3D

As the sharp point of a 3D wedge drives into an object, the object separates along two sides of the wedge; the wedge acts as an inclined plane as the object slides against the wedge's sides. A narrow isosceles triangle serves as a wedge in 2D. A common 3D wedge features two planes meeting at an edge. An analogous 4D wedge would have two hyperplanes meeting at a plane. Just as a 3D wedge drives into an object with a leading edge (where two planes meet at a narrow angle), a 4D wedge would drive into an object with a leading plane (where two hyperplanes would meet at a narrow angle).

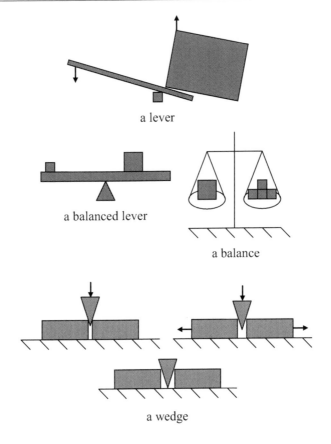

a lever

a balanced lever

a balance

a wedge

A 3D screw can be thought of as a flat strip wrapping around a nail in a helical path; the nail itself is like a right-circular cylinder with a disc-shaped head and right-circular cone at the tip. The 3D screw is a combination of the wedge and the inclined plane: The screw drives into an object like a wedge, and as the turning screw advances the helical strip, the part of the object surrounding the screw wraps around the helical incline. There are two kinds of 4D nails on which to base a 4D screw: A spherinder-shaped nail and a cubinder-shaped nail. A strip could wind around a spherinder-shaped nail to create a 4D screw, while a hyperplane could wind around a cubinder-shaped nail to make another type of 4D screw.

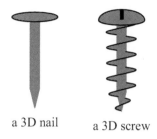

a 3D nail   a 3D screw

The basic function of a gear should be similar in 4D to 3D gears, since gears have a very 2D use even in the known universe (though in 2D it is not possible to have an axle perpendicular to the gear). A gear is basically a wheel with teeth on its rim, where the teeth cause two adjacent gears to interlock such that the rotation of one gear causes the rotation of adjacent gears. The underlying principle of gears is that by rotating one gear of a given size (manually or via a motor), an adjacent gear of different size can be made to rotate at a different angular speed. A similar effect can be achieved with belts and pulleys.

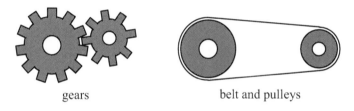

gears   belt and pulleys

A 3D hydraulic press is basically a U-shaped tube filled with fluid (e.g. water), where one column is significantly narrower than the other. The underlying principal is pressure – force per unit area. Less weight on the column with less cross sectional area can be balanced by more weight on the wider column. A 2D hydraulic press would have a similar U-shape with different sized columns, but the fluid would be bound between two curves (rather than by a surface, as in 3D). In 2D, pressure would be force per unit width (rather than area). On the other hand, for a 4D hydraulic press, which would feature a 4D U-shape where the fluid is bound by a hypersurface, pressure would be force per unit volume (the volume of the 3D cross section perpendicular to the vertical axis of the column). This follows from the fact that the volume of a sphere (the cross section of a 4D U-shape with spherinder-shaped columns) is proportional to its radius cubed $(V = 4\pi r^3/3)$, whereas the area of a circle (the cross section of a 3D U-shape with cylindrical columns) is proportional to its radius squared $(A = \pi r^2)$.

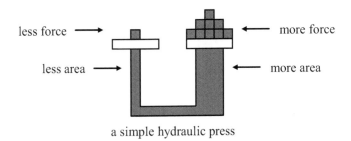

less force ⟶   ⟵ more force

less area ⟶   ⟵ more area

a simple hydraulic press

Puzzle 5.2: Is the hydraulic press more useful in 2D, 3D, or 4D?

The fourth dimension offers extra freedom for connections. In 1D, an object can only have attachments on one of two sides – the front or back. In 2D, there is considerably more freedom for attachments – any direction within the plane. Yet, attachments are still very primitive in 2D, and many common attachments from the known universe do not work in 2D – e.g. nails do not fasten objects together. The third dimension offers the freedom to attach an axle to a wheel without cutting a hole through the rim of the wheel. There would be considerably more room for attachments in 4D compared to 3D. In 4D, an object could be completely enclosed in three independent directions – e.g. it may lie in the center of a cube or a sphere – yet it would still possible to connect it to another object (without putting a hole through the surrounding cube or sphere). For example, the centers of two spheres could be joined by a rod in 4D without cutting a hole in the surface of either sphere – just like the centers of two circles can be joined by a rod in 3D without putting a hole through either circle (the two circles simply need to be parallel). The 3D attachments of the known universe may seem nearly as primitive to a 4D being as 2D attachments seem to a 3D being.

Pipes, tunnels, and holes tend to be shaped like right-circular cylinders in 3D. It would be very challenging to build a non-collapsing pipe or tunnel in 2D because the analogous shape is a pair of parallel lines. The two basic varieties of pipes in 4D would be spherinders and cubinders.

In 3D, one way to connect objects together is to use rope. This requires tying the rope in a knot. It is not possible to tie a rope in a 2D knot, nor would it be possible to tie a rope in a knot in 4D. With the extra degree of freedom in 4D, a knotted rope would come undone easily. However, a rectangular sheet could be knotted in 4D. In 3D, the wrapping of one end of a rope around the rope prevents the knot from coming undone easily, just as the six sides of a prison cell bound a prisoner. However, in 4D, one end of the rope can still travel along the ana/kata direction to untie the knot – as easily as a prisoner could walk along the ana/kata direction to escape from a cubical prison. The ends of a rope have two degrees of freedom in 3D (because one degree of freedom is used up for the length of the rope), while they have three degrees of freedom in 4D. The four edges of a square sheet have one degree of freedom in 3D, but two degrees of freedom in 4D. Two degrees of freedom are needed to be able to tie a knot, while any extra degrees of freedom allow the knot to come undone easily. A square could be knotted in 4D. Just as a line can be extended to a rectangle and a circle extended to a cylinder, a knotted shoelace can be extended to a knotted rectangle in 4D. If a knotted shoelace were extended in 3D, it would intersect itself, but there is no intersection when it is extended along the ana/kata direction in 4D.

In 3D, a one-legged or two-legged stool would tend to fall over, a three-legged stool would balance perfectly, and four or more legs would tend to cause the stool to wobble. The reason is that the floor is a plane in 3D, and any three nonlinear points lie in a plane. A four-legged stool tends to wobble in 3D because it is very difficult to make the fourth leg precisely the right length such that its foot lies in the same plane as the other three feet. Cameras and camcorders, for example, are mounted on a tripod rather than a tetrapod to avoid potential wobbling.

In 2D, where a floor would be a line, a two-legged stool would balance perfectly since any two points are guaranteed to lie in a line, while three or more legs would tend to cause wobbling. A 4D floor would be a hyperplane. Since four nonplanar points define a hyperplane, a three-legged stool would tend to fall over in 4D, while a four-legged stool would be perfectly balanced. In 4D, a camera would be mounted on a tetrapod.

A long, thin rectangle would serve as a good seat for a 2D stool, whereas 3D stools tend to have circular seats with a small thickness. The seat of a 4D stool could be a short, thick spherinder, or it could be a short, thick, wide cubinder, depending upon the shape of a 4D humanoid's bottom. In either case, the seat would sit atop four (or more) legs.

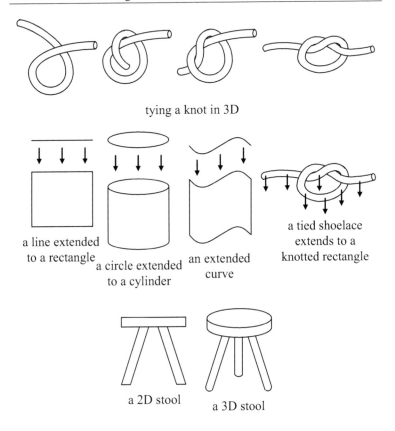

tying a knot in 3D

a line extended to a rectangle

a circle extended to a cylinder

an extended curve

a tied shoelace extends to a knotted rectangle

a 2D stool

a 3D stool

In 3D, no sense of balance is required to ride a tricycle since it has three wheels, while a good sense of balance must be developed to ride a bicycle, and it is even more difficult to ride a unicycle. In 4D, beginning bike riders would first practice pedaling and steering a tetracycle, then move up to a tricycle to develop a good sense of balance. Just as the wheels of a tricycle are not collinear, the wheels of a 4D tetracycle would not be coplanar. A 4D cycle may have four pedals rather than two. In 3D, the two pedals protrude in the left/right directions compared to the plane of the frame of a bike. As the bike is ridden, the pedals rotate in a circle in the forward/backward-upward/downward plane. In 4D, one set of pedals could protrude in the left/right direction and another set in the ana/kata direction, while each set would still rotate in the forward/backward-upward/downward plane. This would be especially convenient for 4D humanoids with four legs.

In 3D, cars have four wheels. Two wheels are connected by an axle, and two such wheel-and-axle combinations are needed to make a planar configuration. In 4D, an analogous four-wheeled car would tip over easily; the wheels must make a 3D configuration in order to be well-balanced on a hyperplanar road. A 4D car could have four wheels, in principle, but not with two axles connecting pairs of wheels with the axles parallel to each other. It would be more natural in 4D to have an eight-wheeled car: There would be a front-right-ana tire, front-right-kata tire, front-left-ana tire, front-left-kata tire, and four analogous rear tires. For cubinder-shaped wheels, there would be four parallel axles, with each axle connecting two parallel wheels. For spherinder-shaped wheels, there would be two sets of parallel left/right axles and two sets of parallel ana/kata axles, with each axle connecting two parallel wheels.

Rooms, closets, cabinets, and drawers would be 4D orthotopes in shape. An orthotope is an N-dimensional rectangular box. A rectangle is a 2D orthotope. The 3D orthotope is called a cuboid (aka rectangular parallelepiped), and looks like a rectangular box with three sets of parallel rectangular faces, where the three types of faces are mutually perpendicular. The 4D orthotope is a rectangular version of the tesseract, bounded by four pairs of parallel cuboids, where the four types of bounding cuboids are mutually perpendicular. The eight cuboid walls of a 4D room would be located on the top, bottom, right, left, front, back, ana, and kata sides of the room. The walls of a 4D room would not be perfect cuboids, but would have a little hyperthickness (just as the walls of a 3D room are not perfect rectangles, but have thickness).

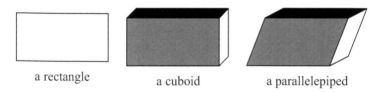

a rectangle   a cuboid   a parallelepiped

In 4D, the door of a cabinet or a room would be a cuboid (with a little hyperthickness), which would hinge about a plane (one side of the cuboid-shaped door) rather than a line (as in 3D) since objects rotate in circles around a plane in 4D (compared to a linear axis in 3D). Doors could swing horizontally in one of four ways in 4D: Hinged at the right side of the cuboid-shaped door, a door would be pulled by a handle at the left side and would rotate in the front/back-left/right plane; hinged at the left side, it would be pulled by a handle on the right side and rotate in the opposite direction of the same plane; hinged at the kata side, the door would be pulled by a handle on the ana side and rotate in the front/back-ana/kata plane; and hinged at the ana side, it would be pulled by a handle on the kata side and rotate in the opposite sense in the same plane. Drawers could slide forward/backward on planar rails in 4D (compared to effectively linear rails in 3D) by pulling on a handle on the front cuboid-shaped side of the drawer.

Each floor of a 2D building would be a linear row of rectangular rooms. In 2D, there would be no hallways unless, for example, the hallways run above or below the rooms. Each floor of a 3D building is a planar arrangement of cuboid-shaped rooms. Hallways run front/back and left/right in 3D. Similarly, each floor of a 4D building would consist of 4D orthotope-shaped rooms in a hyperplanar arrangement. On each 4D floor, hallways would run front/back, left/right, and ana/kata. None of these three directions is uphill or downhill: Every point on any floor is the same distance from the center of the planet. In any N-dimensional space (excluding 0D), the only direction that works with or against gravity is upward/downward.

Zigzag staircases and elevators permit travel up or down to change floors. The stairs are line segments in 2D, rectangular strips in 3D (adjacent stairs sharing edges), and cuboid-shapes in 4D (adjacent stairs sharing planar sides).

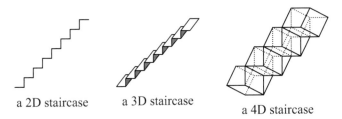

a 2D staircase   a 3D staircase

a 4D staircase

Tables and chairs would have at least four legs in 4D, with more stability resulting from four legs, but more symmetric patterns with paired legs for six or eight legs. An orthotope-shaped table would allow seating at six cuboid-shaped sides in 4D: left, right, front, back, ana, and kata. The top of the table would be cuboid-shaped with a little hyperthickness in 4D, just as an analogous table in 3D has a rectangular top with a little thickness. A desk chair could swivel in two independent directions (or any combination of these) in 4D: rotation in the front/back-left/right plane or the front/back-ana/kata plane.

5.5 Life in NCF4D-U

Now that we have considered both humanoids and structures in 4D, we can put this together to get an idea of what it may be like to live in NCF4D-U.

On the hypersurface of a 4D planet, it would be possible to walk along three independent directions: north/south, east/west, and ana/kata (like a second sort of east/west). In addition, the fourth independent direction – up/down – would be accessible by climbing stairs, a ramp, etc. On a level hypersurface, a 4D level would be balanced if aligned north/south, east/west, or ana/kata. A 4D being would not get closer to or further from the center of the planet by walking along any of these three independent directions (or any combination of them). The ground itself would be 3D – a hyperplane, if the ground were even. Well, locally the ground would look like a hyperplane, but globally it would be a hyperspheroid.

On the surface of a 2D planet, where the two degrees of freedom are forward/backward and up/down, it would be difficult to conceive of a third dimension – left/right – where a whole plane of motion is possible without moving up or down. This is why it is a challenge for 3D beings to contemplate a third degree of freedom on the hypersurface of a 4D planet, in which it is possible to move in a direction perpendicular to north/south and east/west without moving up or down.

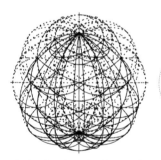

latitudes, longitudes, and hyperlatitudes in 4D

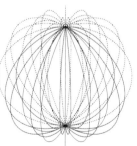

north/south motions

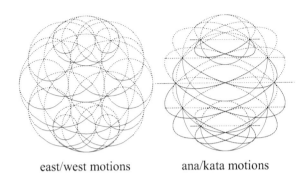

east/west motions   ana/kata motions

In 4D, planets, moons, and raindrops, for example, would be rather hyperspherical. A spinning 4D planet would be shaped more like a hyperspheroid. Hills might curve roughly like the vertex region of a spherical hyperparaboloid in 4D. The basic structure of tree trunks and animal limbs in 4D would be hypercylinders – spherinders or cubinders. Polychora, such as the tesseract, would provide the basis for 4D crystal structures. Artificial objects, constructed by 4D humanoids, would probably reflect basic geometric structures; buildings, rooms, closets, drawers, cabinets, etc. would likely follow an orthotope structure, for example.

In 2D, a line would serve as a road; two tires on a car would both be in contact with the 1D road. In 3D, a road is planar (in particular, a long thin strip); four tires on a car are in contact with the 2D road. In 4D, a road would be hyperplanar (a long, thin, ana-thin 3D strip); the tires (perhaps eight of them) would all be in contact with the 3D road. A 2D town would have essentially a single road, so no roads would intersect (unless, perhaps, one road is above another). Two roads may intersect at right angles in 3D. In 4D, three level roads may intersect at right angles in the same place. Such a traffic signal would feature a three-way light, and right, left, ana, and kata turns would be possible. The usual right-hand rule of math and physics, explained in Chapter 6, may dictate which turns must yield.

A 4D being could experience the full four degrees of motion via sub-hypersurface swimming; swimming along the hypersurface only would be 3D. Boats, birds, and fish, for example, swimming faster than the speed of water waves would create hypercone-shaped shock waves (similar to the V-shaped shock waves that boats and ducks make in rivers), akin to the shock waves created in "air" (which, like "water," would inherently be different in 4D) by objects flying faster than the speed of sound in "air" (which listeners perceive as a sonic boom when it reaches their ears).

A 4D humanoid would see a 3D image on the hypersurface of a 4D eye, and would be able to perceive hyperdepth (the fourth dimension) similarly to how depth (the third dimension) is perceived in the known universe, where a 2D image is cast on the surface of 3D eyes. Technically, a 3D image would be formed by the 4D lens of the eye when viewing a 4D object – the image would not be "on" the hypersurface of the eye; this image would then transferred to the mind, with a method analogous to the rods and cones of human eyes. A 2D humanoid would only be able to see the four lines bounding an opaque square, but none of the points inside the square (and at least two of the four edges would be hidden from sight). A human can see every point inside the six square sides bounding an opaque cube, but points inside the volume of the cube are out of sight (and at least three sides are hidden from a given perspective). A 4D humanoid would be able to see every point inside the eight cubes bounding a tesseract, but none of the points within its hypervolume.

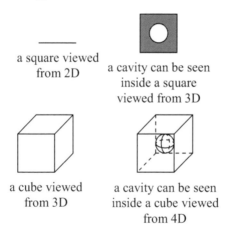

a square viewed from 2D

a cavity can be seen inside a square viewed from 3D

a cube viewed from 3D

a cavity can be seen inside a cube viewed from 4D

Puzzle 5.3: If a human could visit NCF4D-U, would 4D beings be able to see the interior of the human's body – digestive track, lungs, heart, brain, etc.?

A sheet of paper in 3D is planar – basically, a cuboid with tiny thickness, such that the paper is mostly 2D. A sheet of paper in 4D would be hyperplanar, with such a small hyperthickness so as to be effectively 3D. While a sheet of paper in 3D is shaped like a rectangle,

155

a sheet of paper in 4D would be shaped like a cuboid. Thus, it would be possible to paint a 3D picture with a paintbrush without the need of perspective; perspective would instead be used to depict the fourth dimension. The paintbrush would be able to apply paint to every point within the 3D canvas – an impossible feat in 3D. A pair of 4D scissors would be designed to cut the cuboid-shaped paper in half. Instead of the square sticky notes used in the known universe, it may be common to see cube-shaped sticky notes on hypersurfaces in 4D.

If 4D humanoids had 24 fingers – i.e. 4 hands, each with 4 fingers and 2 thumbs, where the extra thumb allows for the extra sense of opposition useful for grasping objects in 4D, and the 4 hands correspond to the number of distinct types of such hands possible in 4D – they would be more likely to adopt a dodecimal number system than a decimal system. The dodecimal number system would have 12 basic digits, including zero. Compare this to the decimal system, which features 10 digits (0 thru 9). Alternatively, the basic number system might include 6 or 24 digits. Such a 4D being could count up to 24 on its fingers without repeating fingers (or being clever, like using one set of fingers to keep track of sixes), and up to 48 if also using toes (if the number of toes equals the number of fingers). Another possibility is a 4D humanoid with two sets of four fingers and one thumb per hand, where the two sets of fingers and thumb are mutually opposable. In this case, they would be more likely to adopt a base-9 number system.

The dodecimal number system is similar to the decimal system, but based on 12, rather than 10, digits; with human symbols, these digits would be 0 thru 9, alpha (α), and beta (β). The Greek symbol α represents the number ten and β represents eleven. In the dodecimal system, ten and eleven are single-digit numbers. The two-digit number 10, which is normally called *ten* in the conventional decimal system of humans, would actually be the twelfth number in a dodecimal system; 18 is the 20th number, 1α is the 22nd number, and 20 is 24th number of the dodecimal system; $6+6=10$; $20-1=1\beta$; etc.

> **Puzzle 5.4**: What do 35, 3α, $\alpha2$, and $\beta\alpha$ in base 12 represent in base 10? What does 35 in base 9 represent in base 10? What do 32 and 100 in base 10 represent in base 12 and base 9? In a popular sci-fi trilogy, $6 \times 9 = 42$. In what base is this true? What trilogy is this?

Pants, shorts, shirts, jackets, etc. for a 4D humanoid may need to accommodate up to four arms and four legs. There would be two waist sizes – one waist size in the front/back-left/right plane and a hyperwaist measurement in the front/back-ana/kata plane; similarly, sleeves would need another measurement aside from the usual length in 3D. Shoe sizes would include length, width, and hyperwidth. A 4D belt may be a long, wide, thin, hyperthin hyperplanar strip – i.e. the two main dimensions (three feet or so, if comparable to the length of a typical 3D belt) would look like a square sheet, then there would be a thickness corresponding to the usual width (an inch or two, typically) of a 3D belt and hyperthickness corresponding to the small thickness (millimeter-sized) of a 3D belt.

Like belts, 4D shoelaces may have two large dimensions – like a square sheet – and two tiny thicknesses. A squarish shoelace could be easily knotted in 4D, in contrast to 3D. Purely 3D shoelaces would easily come untied in 4D, though any physical shoelaces would inherently be 4D structures, even if some of its dimensions were comparatively small. A 50-cm long, 5-cm wide rectangular sheet with a tiny thickness, for example, could be knotted in 3D, but a 50 cm × 50 cm sheet with tiny thickness could not (without folding or cutting it); the object must be roughly 1D to be knotted in 3D (i.e. one length large compared to its other dimensions). However, in 4D, a 50 cm × 50 cm sheet with tiny thickness and tiny hyperthickness could be knotted; a roughly 2D object can be knotted in 4D (two large dimensions and two small ones).

In 3D, there are combs and brushes. A comb is chiefly a 2D structure, consisting of a row of parallel teeth, whereas a brush is fully 3D, with rows and columns of teeth. In 2D, there would just be a comb. In 4D, there would be a comb and two types of brushes, one of which would be a fully 4D structure, with rows, columns, and "ana-columns" of teeth (where ana-columns are perpendicular to both rows and columns), and the teeth are perpendicular to the hyperplane from which they protrude. Whereas 3D hair parts on either side of a line across the surface of the scalp, 4D hair would part on either side of a plane across the hypersurface of the scalp (just like a line can bisect one square bounding a cube, a plane can bisect one cube bounding a tesseract).

Further Reading

If you were engaged by the geometric concepts of the 'toy model' discussed in this chapter, you may enjoy Rudy Rucker's *Spaceland* [A7] – a 4D novel.

Volume 2 Contents

References and Further Reading

These references are divided into two categories: References beginning with an A, such as [A3], are reasonably accessible to a general interest audience; those beginning with a T, as in [T2], are highly technical papers. The technical references have been kept to a minimum and are included primarily to pay tribute to a few researchers who have motivated modern-day experimental searches for large extra dimensions.

Accessible References

A1. *Flatland: A Romance of Many Dimensions,* Edwin A. Abbott, Book Jungle, 2007.

A2. *Flatterland: Like Flatland, Only More So,* Ian Stewart, Perseus, 2001.

A3. *The Fourth Dimension: A Guided Tour of the Higher Universes,* Rudy Rucker, Houghton Mifflin, 1984.

A4. "Large extra dimensions: A new arena for particle physics," N. Arkani-Hamed, S. Dimopoulos, and G.R. Dvali, *Phys. Today* 55N2, 35, 2002.

A5. *Projective Geometry,* H.S.M. Coxeter, Springer, second edition, 2003.

A6. *Regular Polytopes,* H.S.M. Coxeter, Dover, 1973.

A7. *Spaceland: A Novel of the Fourth Dimension,* Rudy Rucker, Tor, 2003.

Technical Papers

T1. "A possible new dimension at a few TeV," I. Antoniadis, *Phys. Lett.* B246, 377, 1990; "New dimensions at a millimeter to a Fermi and superstrings at a TeV," I. Antoniadis, N. Arkani-Hamed, S. Dimopoulos, and G. Dvali, *Phys. Lett.* B436, 257, 1998.

T2. "The hierarchy problem and new dimensions at a millimeter," N. Arkani-Hamed, S. Dimopoulos, and G.R. Dvali, *Phys. Lett.* B429, 263, 1998; "Phenomenology, astrophysics and cosmology of theories with submillimeter dimensions and TeV scale quantum gravity," N. Arkani-Hamed, S. Dimopoulos, and G. Dvali, *Phys. Rev.* D59, 086004, 1999.

Puzzle Solutions

Answers/solutions to selected problems follow.

Puzzle 2.1: The projected images are formed by a right-circular cylinder with its axis parallel to the y-axis.

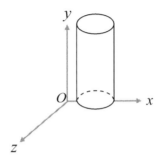

Puzzle 2.2: Calling each tick mark one unit, the center of the glome lies at $(4,3,5,2)$, where the coordinates are given in the order (x,y,z,w).

Puzzle 3.1: The 8 cubes seem to lie on the top and bottom, right and left, front and back, and inside and outside; however, these are really up/down, right/left, front/back, and ana/kata. The "inner" and "outer" cubes are really the same size – in fact, all 8 cubes are congruent. One of the perspective points lies in the center, where 8 parallel (although they do not quite look it) edges meet; the other lies far from the tesseract, up to the right, where 8 more parallel edges meet. The remaining edges include 8 horizontal and 8 vertical edges. There are six sets of parallel faces.

Puzzle 3.2: Yes. Starting with a cube oriented in such as way as to produce a rectangular cross section, a little slanting can produce a cross section in the shape of a trapezoid.

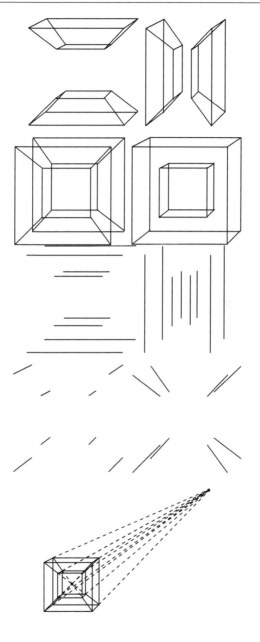

Puzzle 3.3: A body diagonal of Cube 1 is perpendicular to the plane. A face diagonal of Cube 2 is perpendicular to the plane. The front and back faces of Cube 3 are parallel to the plane.

Puzzle 3.4: Initially and finally, the plane intersects the cube through the face diagonals of two parallel sides, for which the cross section is a rectangle. The axis of rotation bisects this rectangle. As the cube rotates about this axis, the cross section becomes a hexagon until the plane intersects at opposite corners of the body diagonal, in which case the cross section is a rhombus. The rhombus shrinks until becoming a square momentarily. The square then grows into a rhombus, reversing the pattern back to a rectangle (corresponding to a 180° rotation).

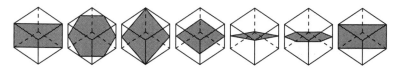

Puzzle 3.5: The arrangements with the following tiles can be folded into a cube: ▨, ■, ◪, and ☰. All together, there are eleven distinct ways to unfold a cube.

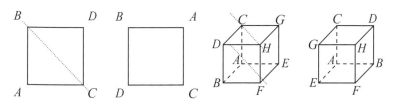

Puzzle 3.6: Yes. Just as a 2D object can be reflected via a rotation in 3D space, a 3D object can be reflected through a rotation in 4D space. For example, consider a square with corners A, B, C, and D, where A and D are opposite corners. In 3D space, this square can rotate about the $\overline{BC}$-axis, for example. If it rotates 180° about the $\overline{BC}$-axis, then A and D swap while B and C remain unchanged. Now consider a cube with corners A, B, C, D, E, F, G, and H, where the first four corners and the last four corners make two parallel faces with A across E, B across F, C across G, and D across H. In 4D space, this cube can rotate about the $ACHF$ plane, for example. For a 180° rotation about the $ACHF$ plane, A, C, H, and F will remain unchanged while D swaps with G and B swaps with E.

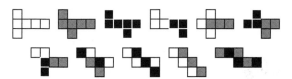

Puzzle 3.7: 35 cubes meet at every corner of a 6D cube.

Puzzle 3.8: The polygonal cross sections of the tetrahedron include the triangle and quadrilateral. The polyhedral cross section of a pentachoron is a tetrahedron.

Puzzle 3.9: It's essentially the same as the hypercube pattern. 252.

Puzzle 3.10: 56.

Puzzle 4.1: Well, its cross section does depend on where the axis of rotation is; it will be a stationary sphere if the axis is through its center, but not otherwise.

Puzzle 4.2: Try hiding the top row of the simplex pattern.

Puzzle 5.1: Left-handed.

Puzzle 5.2: 4D. The extra degree of freedom of force per unit volume, compared to force per unit area or force per unit width, means that the ratio of the output force to the input force will equal the ratio of the radii cubed, compared to squared or times unity.

Puzzle 5.3: Yes, if humans are purely 3D beings. If humans have a slight hyperthickness, like a paper cutout has a small thickness, then no.

Puzzle 5.4: $6 \times 9 = 42$ in base 13; from *The Hitchhiker's Guide to the Galaxy*.

Author's Qualifications

Chris McMullen has published the following journal articles on the collider phenomenology of string-inspired, large extra dimensions:

1. "A Mechanism for Kaluza-Klein number violation in universal extra dimensions," C.D. McMullen and S. Nandi, *J. Phys. G: Nucl. Part. Phys.* 35, 095002, 2008.
2. "Collider implications of a non-universal Higgs," C.D. McMullen and S. Nandi, *Phys. Rev.* D75, 095001, 2007.
3. "Collider implications of multiple non-universal extra dimensions," R. Ghavri, C.D. McMullen, and S. Nandi, *Phys. Rev.* D74, 015012, 2006.
4. "Collider implications of models with extra dimensions," C. Macesanu, C.D. McMullen, and S. Nandi, *Amsterdam, 2002*, ICHEP, 764, 2002.
5. "New signal for universal extra dimensions," C. Macesanu, C.D. McMullen, and S. Nandi, *Phys. Lett.* B546, 253, 2002.
6. "Collider implications of universal extra dimensions," C. Macesanu, C.D. McMullen, and S. Nandi, *Phys. Rev.* D66, 015009, 2002.
7. "Collider implications of Kaluza-Klein excitations of the gluons," D.A. Dicus, C.D. McMullen, and S. Nandi, *Phys. Rev.* D65, 076007, 2002.

The author has taught three intense 30-hour winter courses on the subjects of the fourth dimension and string-inspired extra dimensions at the Louisiana School for Math, Science, and the Arts.

Chris McMullen teaches physics at the Louisiana School for Math, Science, and the Arts – a unique specialized school for high-aptitude, high-achieving students from across the state, where two-thirds of the faculty hold a Ph.D. in their subject area. Students who opt to continue their studies in-state have many opportunities to transfer up to two years of their collegiate course work.

The author is also an adjunct physics instructor for Northwestern State University of Louisiana and began his teaching career at Penn State Altoona.